TECHNICAL
REPORT

Feasibility of Laser Power Transmission to a High-Altitude Unmanned Aerial Vehicle

Richard Mason

Prepared for the United States Air Force

Approved for public release; distribution unlimited

PROJECT AIR FORCE

The research described in this report was sponsored by the United States Air Force under Contract FA7014-06-C-0001. Further information may be obtained from the Strategic Planning Division, Directorate of Plans, Hq USAF.

Library of Congress Cataloging-in-Publication Data

Mason, Richard, 1970-
  Feasibility of laser power transmission to a high-altitude unmanned aerial vehicle / Richard Mason.
      p. cm.
  Includes bibliographical references.
  ISBN 978-0-8330-5135-6 (pbk. : alk. paper)
  1. Drone aircraft. 2. Solar engines. 3. Lasers in aeronautics. 4. Electric power transmission—Technological innovations. I. Title. II. Title: Feasibility of laser power transmission to a high-altitude UAV.

      UG1242.D7.M35 2011
      623.74'69—dc22

                                                                    2011015978

The RAND Corporation is a nonprofit institution that helps improve policy and decisionmaking through research and analysis. RAND's publications do not necessarily reflect the opinions of its research clients and sponsors.

RAND® is a registered trademark.

© Copyright 2011 RAND Corporation

Permission is given to duplicate this document for personal use only, as long as it is unaltered and complete. Copies may not be duplicated for commercial purposes. Unauthorized posting of RAND documents to a non-RAND website is prohibited. RAND documents are protected under copyright law. For information on reprint and linking permissions, please visit the RAND permissions page (http://www.rand.org/publications/permissions.html).

Published 2011 by the RAND Corporation
1776 Main Street, P.O. Box 2138, Santa Monica, CA 90407-2138
1200 South Hayes Street, Arlington, VA 22202-5050
4570 Fifth Avenue, Suite 600, Pittsburgh, PA 15213-2665
RAND URL: http://www.rand.org/
To order RAND documents or to obtain additional information, contact
Distribution Services: Telephone: (310) 451-7002;
Fax: (310) 451-6915; Email: order@rand.org

Preface

High-altitude, long-endurance aircraft have long been a topic of interest for the U.S. Air Force and for the RAND Corporation. In 2007, RAND undertook a study to determine whether there was any practical merit to the admittedly speculative concept of transmitting power to such an aircraft by photovoltaic conversion of a laser beam, thus recharging the aircraft in midair and potentially keeping it aloft indefinitely.

This report describes work conducted between 2007 and 2010. It was supported by internal research funding from the RAND Corporation and from RAND Project AIR FORCE.

RAND Independent Research and Development (IR&D)

This work was performed within the RAND Corporation's continuing program of self-initiated research. Support for such research is provided, in part, by donors and by the independent research and development provisions of RAND's contracts for the operation of its U.S. Department of Defense federally funded research and development centers. The research described here was administered by the Force Modernization and Employment Program within RAND Project AIR FORCE.

RAND Project AIR FORCE

RAND Project AIR FORCE (PAF), a division of the RAND Corporation, is the U.S. Air Force's federally funded research and development center for studies and analyses. PAF provides the Air Force with independent analyses of policy alternatives affecting the development, employment, combat readiness, and support of current and future aerospace forces. Research is performed in four programs: Force Modernization and Employment; Manpower, Personnel, and Training; Resource Management; and Strategy and Doctrine.

Additional information about PAF is available on our website:
http://www.rand.org/paf/

Contents

Figures

Tables

Summary

Solar-powered unmanned aerial vehicles (UAVs) that have been developed and flown have demonstrated interesting capabilities for high altitude and long endurance. However, current solar-powered UAVs are extremely light and fragile and have small payloads. The concept of a UAV with photovoltaic (PV) cells powered by a laser beam has been demonstrated by the National Aeronautics and Space Administration (NASA) on a tiny scale but has not been applied to a UAV of sufficient size to be of any practical interest.

This report examines whether the laser-beam-powered UAV concept could be scaled up to a practical high-altitude UAV and identifies some of the concept's limiting factors.

The finding of the report is that the concept does have merit, at least in the narrow sense that it is technologically feasible, and it could be used to build a UAV with performance characteristics that are beyond the performance envelope of existing air vehicles, especially sustained extremely high altitude. Commercially available lasers and PV cells could provide a UAV with twice as much power as that of a similar solar-powered UAV. Moreover, the laser can be more consistently available than the sun, thus reducing the need for batteries on the UAV. Even under conservative assumptions, a laser-powered UAV could have four times the payload and 80 percent higher nighttime altitude than a solar-powered UAV of the same size and total weight. With more aggressive assumptions and state-of-the-art lasers and PV cells (demonstrated, but not necessarily commercially available off the shelf), it is possible to achieve up to 10 times the power of a similar solar-powered UAV. Beyond that power level, the concept runs into thermal limits of current state-of-the-art PV cells.

One disadvantage of the concept is that if the laser is beamed from the ground or from a ship, the UAV is closely "tethered" to the beam source and (to receive useful amounts of power) must fly in an orbit within a few tens of kilometers of it. It could, however, fly at extremely high altitudes over the beam source. Another problem with the concept is that clouds could interrupt the beam and force the UAV to descend below the cloud layer from time to time.

Both of these problems could be circumvented by placing the laser on a conventional aircraft, so that the UAV would be powered by an air-to-air transmission. In this case, the "tether" from the UAV to the power source could be much longer (hundreds of kilometers), and clouds would no longer be a likely threat. Deploying the laser source on an aircraft should be technologically feasible, although flying that aircraft, of course, imposes an additional operational burden.

This report focuses on the physical parameters of flight—altitude, range, persistence, and power—that are possible for a laser-PV aircraft that uses current technology. Whether the performance niche opened by this concept is really valuable and worth pursuing or merely a technology "stunt" waiting to happen is debatable (and could be examined in a future study). Jet propulsion is generally a superior technology, except for missions requiring extreme endurance or extremely high altitude. Because of the effort required to support a laser-PV UAV above the cloud layer, the "extreme endurance" argument is not very compelling. However, the laser-PV concept could be worth further consideration if an important mission were identified for an air vehicle with ultra-high operating altitude and moderate persistence and payload. Some possibilities include ultra-high-altitude observation stations or communication relays and flocks of high-altitude sensor probes powered remotely from a large aircraft "mother ship."

Acknowledgments

I would like to thank Jim Thomson, Andrew Hoehn, and Richard Moore for their support and encouragement of this research. I thank John Tonkinson for guidance on flight performance and Paul DeLuca for useful discussions. Jon Grossman and John Friel each provided useful information on laser systems. Chad Ohlandt and Eric Johnson reviewed the document and made many helpful criticisms and suggestions. Any remaining deficiencies, however, are the responsibility of the author.

Abbreviations

ABL	airborne laser
CIS	copper indium selenide
COIL	chemical oxygen iodine laser
DARPA	Defense Advanced Research Projects Agency
DC	direct current
DPAL	diode-pumped alkali laser
DPSSL	diode-pumped solid-state laser
GaAs	gallium arsenide
GEO	geostationary Earth orbit
InGaAs	indium gallium arsenide
kg	kilogram
kgf	kilogram-force
kJ	kilojoule
kW	kilowatt
MW	megawatt
NASA	National Aeronautics and Space Administration
nm	nanometer
PAF	Project AIR FORCE
PV	photovoltaic
UAV	unmanned aerial vehicle

CHAPTER ONE

Background

The Value of Persistence at High Altitude

The effectiveness of an aircraft or spacecraft as a platform for various sensor and communications payloads depends on its ability to attain and maintain a line of sight to the target to be sensed or communicated with. The higher the operating altitude of an aircraft or spacecraft, the greater the area of the Earth's surface to which it has a direct line of sight at greater slant ranges for any given zenith angle; thus, at very high altitudes, the platform can communicate across a greater area or sense a target at a greater standoff distance. However, greater altitudes are not unequivocally good—if the platform is higher than necessary, the slant range will be longer than necessary, typically reducing the performance of the sensor or communications link. For any particular mission, there is an ideal altitude that balances these considerations.

Satellites, of course, operate at very high altitudes. Furthermore, orbital mechanics decrees that a satellite in a relatively low orbit will pass over a given point quickly, and extremely high orbits are required to keep a particular point on the Earth continuously in view for extended periods of time. Figure 1.1 illustrates the most favorable case for persistence (of visual line of

Figure 1.1
Required Satellite Altitude for a Given Persistence

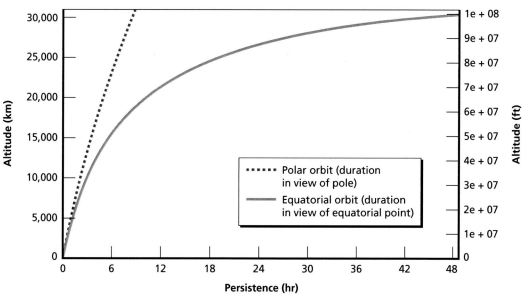

sight at positive elevation), in which the ground site is on the equator and the satellite is in an equatorial orbit. Figure 1.1 also illustrates the case where the ground site is at the pole and the satellite is in a polar orbit. In either case, a persistence of hours requires a satellite altitude of thousands of kilometers.

Satellites are expensive and not generally available for launch on short notice. Once they are launched, their orbits cannot easily be changed and can be predicted well in advance by an adversary. Aircraft, on the other hand, are responsive and can be persistent at much lower altitudes. There is, therefore, a potential role for aircraft that attain the highest altitudes and greatest endurance possible, while still being cheaper, more responsive, and a few orders of magnitude closer to the ground than satellites are.

The radius of the ground circle visible to an aircraft at a given altitude is plotted in Figure 1.2. For example, an ultra-high-altitude aircraft could observe significant areas of mainland China while orbiting over Taiwan, as illustrated in Figure 1.3. (Figure 1.3 shows the visible horizon; it does not show shadows due to terrain masking, but avoiding these is another benefit of very high altitude.) Alternatively, an airborne platform at 100,000-ft altitude could operate as a communications relay across a country the size of Iraq. The latency associated with light-speed transmission to such a relay would be only a few milliseconds, whereas if a geostationary communications satellite was used for the same purpose, a quarter-second of latency would be introduced every time a signal traveled to geostationary Earth orbit (GEO) and back.

Photovoltaic Aircraft

In the past decade, some aircraft that use solar photovoltaic (PV) cells to drive electric motors have been built. These aircraft have the potential to operate at altitudes at which conventional

Figure 1.2
Radius of the Ground Circle with Line of Sight to an Airborne Platform, as a Function of the Platform Altitude and the Minimum Required Elevation Angle of the Line of Sight

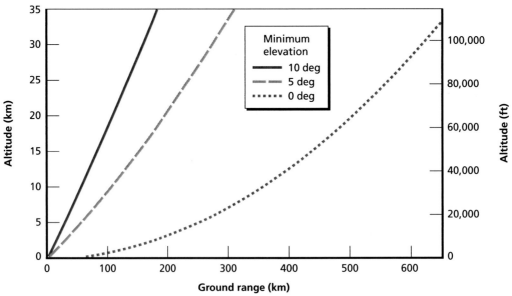

Figure 1.3
Region Visible to Surveillance Aircraft Flying at Increasing Altitudes over Southeastern Taiwan

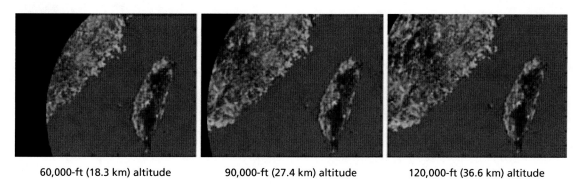

| 60,000-ft (18.3 km) altitude | 90,000-ft (27.4 km) altitude | 120,000-ft (36.6 km) altitude |

RAND *TR898-1.3*

air-breathing engines might be starved of oxygen. Figure 1.4 plots the operational limits of several high-altitude aircraft. In general, solar-PV aircraft are capable of higher altitude and longer endurance than air-breathing aircraft. (Figure 1.4 does not reflect the possibility of aerial refueling, which could, of course, enable conventional aircraft to remain aloft for longer times.) Solar-PV aircraft have demonstrated both record-breaking altitudes and record-breaking endurance, though not yet at the same time. With enough electrical storage to fly through the night and recharge again the next day, solar-PV aircraft might be able to fly indefinitely without the need to refuel.

The AeroVironment Helios Prototype unmanned aerial vehicle (UAV) had a 75-m wingspan, a takeoff weight of 719 kg, and solar cells with a nominal total power of 31 kW, powering 14 1.5-kW motors. On August 13, 2001, the HP01 version of the Helios Prototype flew for more than 40 minutes at 29,524 m, setting a new altitude record for sustained flight by a

Figure 1.4
Altitude and Unrefueled Endurance Limits for Various Aircraft

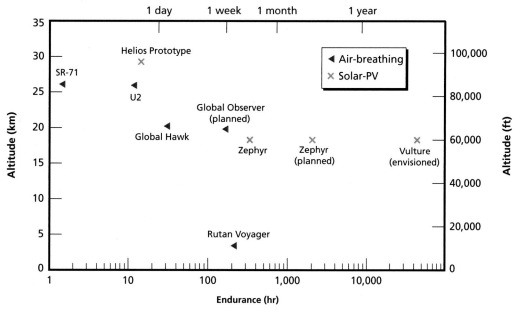

RAND *TR898-1.4*

winged aircraft. This version was more or less daylight-limited and was estimated to have an endurance of 14 to 15 hours.

Two years later, a "high-endurance" configuration of the Helios Prototype (HP03) was ready for testing. It had a gross weight of 1,052 kg, 10 motors instead of 14, and a rechargeable hydrogen-air fuel cell. The HP03 was intended to have an endurance of 7 to 14 days. However, before this endurance could be demonstrated, on the aircraft's second test flight, on June 26, 2003, the aircraft encountered turbulence, which caused the wing to deform into a high-dihedral-angle configuration. The design of the aircraft made it slow to return to a nominal dihedral angle, and in the persistent high-dihedral configuration, it was unstable in pitch. The unstable pitch oscillations caused the aircraft to break apart in midair (Noll et al., 2004), as illustrated in Figure 1.5.

On June 1–3, 2005, a smaller solar-powered UAV, the AC Propulsion SoLong, flew for 48 hours. The SoLong has a 4.75-m wingspan, 12.8-kg takeoff weight, and 225 W of nominal solar panel power. It consumes 95 W in level flight, and its maximum motor power is 800 W (Cocconi, 2005). The SoLong's pilots used gliding and soaring tactics to exploit desert thermals and conserve power. The proximate reason for landing the SoLong after 48 hours and 16 minutes was pilot exhaustion.

Zephyr, a solar-powered UAV with an 18-m wingspan and a takeoff weight of 30 kg was developed by the British company QinetiQ, with funding from the UK Ministry of Defence and the U.S. Department of Defense as a Joint Capability Technology Demonstration. Zephyr uses amorphous silicon solar arrays with a nominal power of 1.5–2.0 kW during daylight, charging lithium sulfur batteries for use in darkness. On September 10, 2007, Zephyr made a 54-hour flight, attaining a maximum altitude of 17,786 m (Noth, 2008b). On August 24, 2008, Zephyr made a flight of 82.6 hours, with a maximum altitude of over 60,000 ft (18,288 m) (Millard, 2008). On July 23, 2010, a larger version of Zephyr with a 22.5-m wingspan and 52-kg weight completed a flight of 14 days and 24 minutes, crushing the previous duration record for unmanned flight[1] (Page, 2010). The SoLong and Zephyr are shown in Figure 1.6.

Figure 1.5
Flight of Helios HP03 and Its Destruction by Unstable Pitch Oscillations Excited by Turbulence

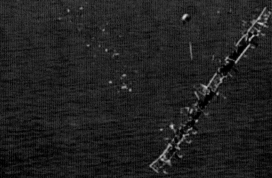

Helios Prototype UAV   Helios Prototype breaks up in midair

RAND TR898-1.5

---

[1] The previous official Fédération Aéronautique Internationale world duration record for a UAV—30 hours, 24 minutes—was held by Global Hawk. The Zephyr flights prior to 2010 were not officially considered world records because no Fédération official was present.

Figure 1.6
Small to Midsize PV UAVs

AC Propulsion SoLong   QinetiQ Zephyr

RAND *TR898-1.6*

The Defense Advanced Research Projects Agency (DARPA) is funding the Vulture program, which envisions an unmanned aerial system with a 1,000-lb payload, drawing 5 kW of power, that is able to stay continuously airborne for at least five years. Three contractors—Aurora Flight Sciences, Boeing, and Lockheed Martin—performed analytical concept studies in the first phase of the program (Walker, 2008). All three Vulture concepts were PV aircraft (Figure 1.7). Phase II of the program, now under way, is intended to produce a demonstrator aircraft capable of at least 30 days of continuous flight with a 200-lb payload, drawing 1 kW of power (DARPA, 2009).

Concept for a Useful Laser-Powered UAV

All of the PV aircraft described above rely on sunlight as their power source, and indeed, sunlight has many advantages: It is free, widely available, fairly intense, and immensely powerful across its total beam. However, in some respects, the sun falls short of being an ideal power source. Most obviously, it is not available at night. What if an artificial light source were used instead? To investigate this possibility, this report considers PV aircraft with a laser serving as the illuminating power source.

Figure 1.7
Artist's Rendering of Three Concepts in the DARPA Vulture Program

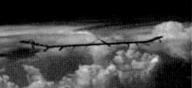

Aurora Flight Sciences   Boeing   Lockheed Martin

RAND *TR898-1.7*

Assume that the laser-powered aircraft under discussion are UAVs. Since the intent of PV aircraft is to achieve high altitude and long persistence, albeit with a small payload, a human pilot on board would probably be only a liability. By definition, laser-powered aircraft fly within the line of sight of some power-supplying unit which could also supply command and control.

For this analysis, also assume that the laser beam originates from a ground site or from a ship or aircraft operating at a significantly lower altitude than the UAV itself. It is possible to imagine the laser beam arriving at the UAV from above, perhaps bounced down from a satellite or airship, but this seems like an excessive complication of the concept. If it were possible to place some platforms overhead that could reliably and persistently power the UAV, those platforms could themselves fill the persistent high-altitude niche we are trying to address with the laser-PV UAV.

For a laser-powered UAV to be attractive, the laser should deliver power at least comparable to that available from solar power (after allowing for differences in conversion efficiency). At lower levels of power, the laser would be, at most, a minor supplement to a solar-PV aircraft. Therefore, the laser's irradiance at the aircraft should be 1000 W/m^2 or more, and the total power delivered should be at least a few kilowatts (comparable to that of Zephyr) or tens of kilowatts (comparable to that of Helios).

The mass of a solar aircraft intended to fly through the night consists primarily of batteries, followed by structural mass (Noth, Siegwart, and Engel, 2006). For example, batteries made up 44 percent of the total mass of the AC Propulsion SoLong (Cocconi, 2005). The hydrogen fuel cell and hydrogen tanks that enabled the Helios Prototype to fly through the night made up 37 percent of its total mass (Noll et al., 2004). If a laser can approximately replace sunlight, the PV aircraft can shed much of the energy-storage mass that it would otherwise need to fly through the night. If the storage mass were reduced by, say, 80 percent (so that the UAV had reserves to fly for 2 hr without laser illumination, instead of 10 hr in darkness), the UAV's payload fraction would be dramatically increased.

If the laser could deliver higher irradiance than sunlight, the UAV could be designed to be more capable in flight or to carry a higher-power payload. At the extreme, some existing PV cells are capable of usefully absorbing irradiances on the order of 500,000 W/m^2 (Spectrolab, 2008). Multiple lasers could be separately focused on the UAV to produce a total combined irradiance greater than that of a single laser. Such a "modular" approach could be attractive, e.g., by allowing incremental improvement or graceful degradation as individual laser modules went on or off line. However, the benefits of modularity must be balanced against the cost of a separate telescope and beam director for each laser (Kare, 2004).

Some aspects of the system to consider are the generation of laser light at the power source, with the attendant local cooling requirement; the optics necessary to direct the beam at the UAV; the attenuation of the beam as it passes through the atmosphere, including possible obstruction by clouds; the efficient absorption of the beam by PV cells on the UAV and thermal management of the PV cells; and finally, the translation of that power into useful thrust by electric motors.

The laser should be able to power the UAV at a slant range of at least 40 km. (With this range, the UAV could achieve extreme altitudes while orbiting close to the laser source.)

Advantages of an Artificial Light Source

The amount of daylight in a 24-hour period varies as a function of latitude and season, but on average, the sun is available only about half of the time. When the sun is present, it does not remain at a consistent place in the sky; it moves over the course of the day, so it will seldom be at the optimal angle for collection by a solar aircraft, unless the aircraft continually alters its orientation to track the sun. Since aircraft generally have more horizontal than vertical surface area, and existing PV aircraft place their solar arrays on the wings, it would be ideal to have the sun always overhead in perpetual high noon.

Angle and Availability

An artificial light source such as a laser could, in principle, operate at any time of the day or night. With coordinated choices of aircraft design, laser location, and aircraft flight path, the angle between the laser beam and the collecting PV cells could be good almost all the time. In particular, if the UAV were intended to remain orbiting above the laser source (possibly as a high-altitude observation post or communications relay), the beam could always be nearly normal to the aircraft's wing surfaces, for ideal collection geometry. If the laser source were itself placed on a mobile platform, the UAV could move with it.

If the UAV were intended to fly far from the laser source so that the laser beam would be closer to horizontal, continually presenting a PV collection surface to the beam would be more complicated. However, it could be achieved most of the time with the right choice of flight path, e.g., a "racetrack" orbit orthogonal to the beam. Alternatively, two laser sources could be available from different directions, with the UAV handed off from one to the other when it changed heading.

Light Frequency

The "bandgap energy" of a PV material is the energy necessary to free an electron in the material from its covalent bond. Since the energy of a photon is proportional to its frequency, PV cells are responsive to particular frequencies of light corresponding to the cell's bandgap energies. Photons at too low frequency do not generate electric current, whereas high-frequency photons waste energy in excess of the bandgap energy. The spectral-response curves for some different PV materials are illustrated in Figure 1.8 (Krupke et al., 2003). The ideal light source would be monochromatic and at an ideal frequency for the PV collector material, instead of spread across a broad spectrum of frequencies, as sunlight is.

Light Intensity

Finally, the power flux of direct solar radiation is roughly 1,350 W/m^2 at the top of the atmosphere and roughly 1,000 W/m^2 at sea level. However, PV cells in current production can usefully absorb as much as 1,000 sun-equivalents (with 500 sun-equivalents being a recommended operating level). These cells are used to generate power from sunlight that has been concentrated by lenses by a factor of 500. Irradiated with more intense light, PV cells not only generate more absolute power, they are also more efficient for a given temperature, as illustrated in Figure 1.9 (Spectrolab, 2008). Thus, a PV aircraft could generate much more power for a given array area with a light flux of 500,000 W/m^2, assuming that the PV array on the aircraft could be kept adequately cool.

Figure 1.8
Spectral Response of Some PV Materials

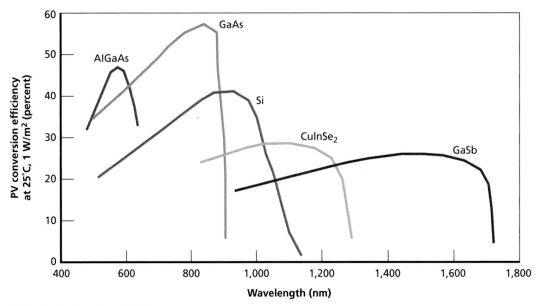

SOURCE: Krupke et al. (2003).
RAND *TR898-1.8*

Figure 1.9
Efficiency of Spectrolab CDO-100 Solar Cells, as a Function of Temperature and Incident Light Intensity

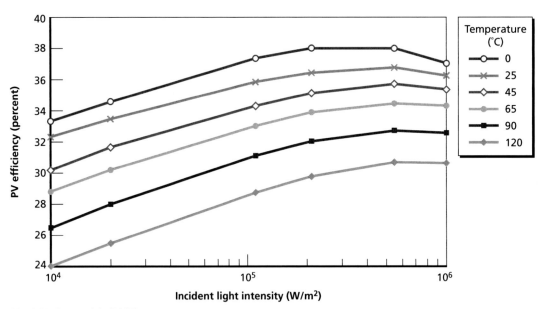

SOURCE: Spectrolab (2008).
RAND *TR898-1.9*

It is possible, therefore, that an artificial light source could be a better power source than the sun for a PV aircraft. A laser beaming power to the PV aircraft could, in principle, operate 24 hours per day. Its location would be consistent and/or controllable. With the right choice of flight path, the beam could be kept nearly normal to the aircraft's PV collection surfaces for good collection efficiency. The laser and the PV array on the aircraft could be tailored to a mutual frequency for high efficiency. Finally, if the laser were bright enough, the total power flux to the aircraft could be greater than that which would be possible with sunlight.

The properties of an ideal artificial light source are compared with those of sunlight in Table 1.1.

Table 1.1
Properties of the Sun and Those of an Ideal Light Source for PV Aircraft

Property	The Sun	Ideal Light Source
Duty cycle	Approximately 50% (unavailable at night)	100%
Intensity	Approximately 1,000 W/m^2	500,000 W/m^2
Wavelength	Broad spectrum	Narrow band tailored to PV material: 800 nm for GaAs
Angle to light source	Changes with time of day, time of year, latitude	Normal to collecting surface area of aircraft

CHAPTER TWO

Laser Power Beaming to a Photovoltaic Cell

A small demonstration of laser power beaming to a UAV was performed by the National Aeronautics and Space Administration's (NASA's) Dryden Flight Research Center and Marshall Space Flight Center in 2003 (Blackwell, 2004). The UAVs, named MOTH1 and MOTH2, were small radio-controlled planes weighing 9 or 10 oz with a 2-m wingspan. The MOTH2 is shown in Figure 2.1. The laser chosen by the Marshall Space Flight Center engineers was a 1.5-kW 940-nanometer (nm) diode array with high (50 percent) efficiency. For the PV cells, they chose 17 percent efficient silicon cells as a cost-effective match to the laser frequency.

The laser beam at 15.2-m range was 85 to 95 cm in diameter, resulting in an irradiance of 560 W/m^2, delivering 40 W of irradiance to the PV array and 7 W of power to the propeller motor. The beam spot size was rather large compared with that of the PV array, and this seems to have been the greatest single inefficiency in the power transmission (greater than losses in the laser, PV cells, or electric motor).

MOTH2 made several indoor flights of more than 15 minutes each. A human operator manually tracked the plane with the laser beam, and the limiting factor was operator fatigue; an autotracker might have powered the plane indefinitely.

Figure 2.1
First Flight of the MOTH2 UAV, with Laser Light Illuminating Its PV Array

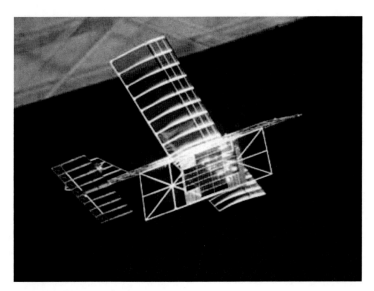

SOURCE: NASA.
RAND *TR898-2.1*

The experimenters then attempted outdoor tests at Redstone Arsenal, but these were not successful because of the difficulty of seeing the laser spot to target the aircraft and inadequate motor power for the aircraft to cope with gusty winds. The MOTH2 aircraft had not necessarily been designed with outdoor tests in mind.

Subsequently, the NASA project lost momentum with the retirement of several of the individuals involved.[1]

Power transmission by laser beam has also been commercially realized on a very small scale. JDSU markets photonic power modules that transmit 0.5 to 1.0 W of power by sending laser light through more than 0.5 km of optical fiber to be converted by a PV cell at the other end (JDSU, n.d.). Such devices are useful for delivering small amounts of power to sensors in environments where electrical transmission is problematic, such as in the vicinity of large magnetic fields, as inside power transformers or magnetic resonance imaging machines (Hecht, 2006). These power modules match the frequency of the laser to the PV cell to achieve efficiency of 40 to 50 percent.

Scaling Up to a Practical High-Altitude UAV

The flux ϕ delivered to the aircraft at slant range L is given by (Kare, 2004)

$$\phi = \frac{R_{\text{source}} A_{\text{source}} \eta_{\text{trans}}}{L^2}$$

(2.1)

where R_{source} is the radiance (power per unit area per steradian) of the laser source, a constant that cannot be changed by passive optics; A_{source} is the total area of the beam source, possibly spread across more than one telescope aperture; and η_{trans} is the transmission efficiency through the atmosphere. The radiance of a coherent beam is

$$R = \frac{P}{\lambda^2 B_x B_y}$$

(2.2)

where P is the beam power, λ is the laser wavelength, and B is a dimensionless number indicating the beam quality (B_x and B_y if the beam quality is different on the two transverse axes).

If the smallest system that might have practical interest requires $\phi = 1{,}000 \text{ W/m}^2$ and $L = 40$ km, it follows that we require

$$A_{\text{source}} R_{\text{source}} > 1.6 \times 10^{12} \text{ W/sr}$$

(2.3)

The largest telescopes in the world are about 10 m in diameter, so $A_{\text{source}} \approx 75 \text{ m}^2$. More readily available telescopes would have apertures up to 1 m in diameter ($A_{\text{source}} \approx 0.8 \text{ m}^2$). This constrains the types of laser that could supply adequate radiance R_{source}. For example, diode or semiconductor lasers are not bright enough to meet this requirement with an aperture of reasonable size; thus, the diode laser array used in NASA's MOTH2 experiment would not scale up to a UAV of practical size.

---

[1]   Tim Blackwell, personal communication, 2007.

Available Types of Laser

Various types of laser that could be considered candidates for beaming tens of kilowatts are listed in Table 2.1 and discussed further in Appendix A. Of these, diode-pumped solid-state lasers (DPSSLs) such as fiber and disk lasers are the most promising existing technology to power a practical laser-PV UAV. Other candidates either are not bright enough, have not yet been demonstrated at high power levels, are environmentally hazardous, or are too far into the infrared for good PV conversion.

Fiber lasers with up to 50 kW of power and greater than 25 percent efficiency are available commercially and are in widespread industrial use for cutting and welding (IPG, 2008). Beam quality is good for single fibers with 2 kW of power and not as good for higher-power fiber lasers; however, commercially available fiber lasers appear adequately powerful and bright for a laser-PV UAV.

Thin disk lasers with 8 kW of power (such as the one in Figure 2.2) are commercially available (TRUMPF, 2008). Numerical models show that power output of more than 40 kW from a single disk is possible (Giesen and Speiser, 2007). Thin disk lasers can achieve very high beam quality and also appear potentially adequate for a laser-PV UAV. Thin disk lasers may be easier than fiber lasers to scale to 100 kW or more while maintaining good beam quality. In June 2008, Boeing fired a thin disk laser with more than 25 kW of power and a beam quality "suitable for a tactical weapon system" for multisecond durations (Selinger, 2008).

As will be shown below, DPSSLs can generate beams with from one to 10 times the intensity of sunlight at useful distances. At higher intensities, the PV panels on a UAV tend to overheat, so the greater brightness of chemical lasers is not very useful, particularly in view of the operational complications of operating a high-power chemical laser (e.g., hazardous chemical exhaust).

Table 2.1
Currently Available Laser Technologies

Laser Type	Wavelength, λ (nm)	Efficiency, η	Beam Quality, $B_x \times B_y$	Radiance, R_{source} (W/m²–sr)
Diode, 10 kW	850	50% DC to light	100 × 1.5	1 × 10^{10}
Spectral beam combining, 25 W	850	25% DC to light	4.5 × 3	2 × 10^{12}
Thin disk, 8 kW	1,060	25% DC to light	24 × 24	1.2 × 10^{13}
Thin disk, 25 kW	1,060	25% DC to light	3 × 3	2.4 × 10^{15}
Fiber, 2 kW	1,060	25% DC to light	1.2 × 1.2	1.2 × 10^{15}
Fiber, 10 kW	1,060	25% DC to light	15 × 15	4 × 10^{13}
Fiber, 20 kW	1,060	25% DC to light	35 × 35	1.4 × 10^{13}
Diode-pumped alkali, 48 W	795	25%–40% DC to light	1.2 × 1.2	6 × 10^{15}
Chemical oxygen iodine, 1 MW	1,315	300 kJ/kg	1.3 × 1.3	3 × 10^{17}
Hydrogen fluoride, 1 MW	2,700	150 kJ/kg	2 × 2	3 × 10^{16}
Deuterium fluoride, 1 MW	3,800	150 kJ/kg	2 × 2	2 × 10^{16}

SOURCES: Kare (2004), TRUMPF (2008), IPG (2008), Zhdanov, Sell,and Knize (2008), Cathcart (2007).

Figure 2.2
An 8-kW Laser Based on Four Thin Disk Laser Heads

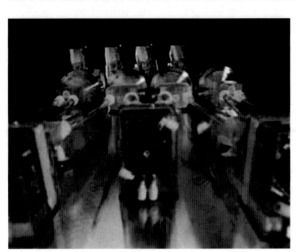

SOURCE: TRUMPF Group (2008).
RAND TR898-2.2

The DPSSLs have the disadvantage that they produce light at wavelengths greater than 1 μm. As shown in Figure 1.8, this is too long for efficient conversion by silicon or gallium arsenide (GaAs) PV cells. However, other PV materials such as germanium or copper indium selenide (CIS) have a spectral response that extends to the near-infrared and can be used to convert the beam to electricity, though perhaps at some penalty in efficiency.

Figure 2.3 shows the conversion efficiency of two PV materials in the near-infrared region of λ = 1,060 nm. CIS films are commercially available and should have conversion efficiency of from 17 to 20 percent, as shown. The conversion efficiency increases for increasing laser intensity, ϕ, but decreases at higher temperatures.

Figure 2.3
Conversion Efficiency of CIS and Tailored InGaAs at λ=1,060 nm

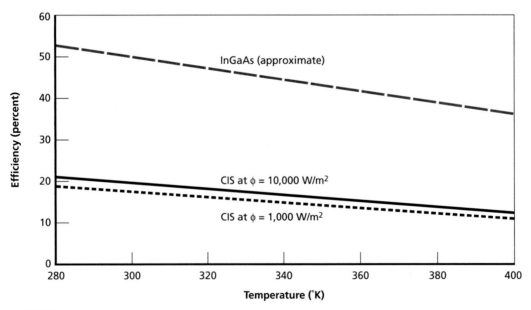

RAND TR898-2.3

Indium gallium arsenide (InGaAs) PV cells could also be tailored to convert near-infrared light by adjusting the relative quantities of indium and gallium. Although such cells are not in commercial production, officials at Spectrolab have stated that they could design an InGaAs device around the λ = 1,060-nm wavelength with an efficiency of about 50 percent.[2] The approximate efficiency of these InGaAs cells as a function of temperature is also estimated in Figure 2.3. The variation of InGaAs efficiency with laser intensity is not shown, since it is probably within the uncertainty of the approximation at a given temperature. PV conversion efficiency is discussed further in Appendix B.

---

[2] Dmitri Krut (Spectrolab), personal communication, 2008.

CHAPTER THREE

Laser Power Transmission Through the Atmosphere

As a laser beam passes through the atmosphere, several effects combine to reduce the amount of energy that actually arrives at the UAV:

- Particles in the lower atmosphere, such as dust, smoke, and water droplets, scatter the beam. This is significant at lower altitudes and can be even more significant if weather conditions are hazy.
- Gas molecules in the atmosphere absorb energy from the beam, but this effect is generally less significant.
- The laser beam spreads due to diffraction; spreading is greater for lasers with lower beam quality.
- Turbulence in the atmosphere causes further spreading of the beam. Again, this is more significant at low altitudes.
- Shaking or vibration of the laser causes the beam to "jitter."
- "Thermal blooming," an optical effect resulting from the laser heating the atmosphere, causes further beam divergence.

The equations governing these effects and their relative magnitudes for fiber and thin disk lasers are described in Appendix C.

Commercially available high-power lasers generally have mediocre beam quality; for them, diffraction and scattering are the dominant effects. Weapon-quality lasers have high enough beam quality that beam jitter and turbulence can also become significant. Thermal blooming is not particularly significant for lasers at the power levels under consideration here.

If the laser is airborne rather than ground-based, beam jitter will possibly be worse, but atmospheric scattering and divergence caused by turbulence will be decreased.

Results of Beam Propagation from a Ground Laser

Figure 3.1 illustrates the flux delivered by commercially available ground-based fiber lasers in both clear and hazy conditions. The limiting factors on the intensity of the flux are particulate scattering of the beam in the lower atmosphere and the beam quality of the laser.

The fact that the laser is more heavily attenuated in the lower atmosphere means that there is a roughly "cone-shaped" volume over the laser ground site in which the UAV can receive a given power flux. The UAV must remain within a few tens of kilometers of ground

Figure 3.1
Flux Delivered by Commercially Available Ground-Based Fiber Lasers with a 1-m Telescope and 1,060-nm Wavelength (kW/m²)

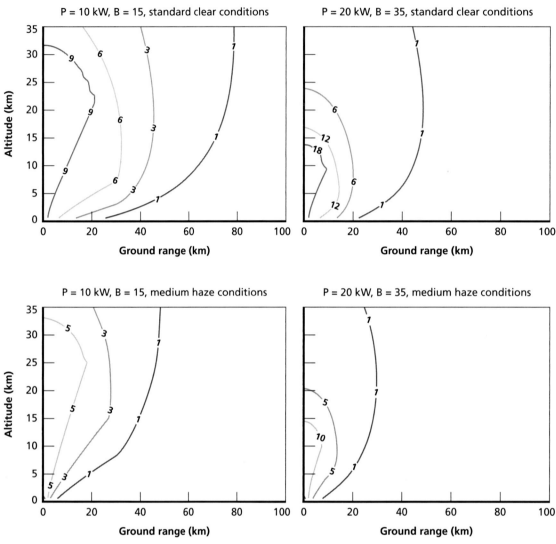

range from the laser to receive a flux that is larger than that of sunlight. This limits the possible operations of the UAV; however, there may be applications for a UAV in a "stationary" high-altitude orbit but closely tethered to a ground source.

Commercially available fiber lasers generally have poorer beam quality at high powers; as shown in Figure 3.1, the 20-kW fiber laser actually delivered worse results because the higher power did not compensate for the poorer beam quality.

Figure 3.2 shows the effect of replacing the fiber laser with a thin disk laser representing the current state of the art (but not necessarily commercially available). The operating volume of the UAV is increased, although the vehicle will still have to operate less than 100 km from the laser source, especially in hazy conditions.

Figure 3.2
Flux Delivered by Current State-of-the-Art Ground-Based Thin Disk Lasers with a 1-m Telescope and 1,060-nm Wavelength (kW/m²)

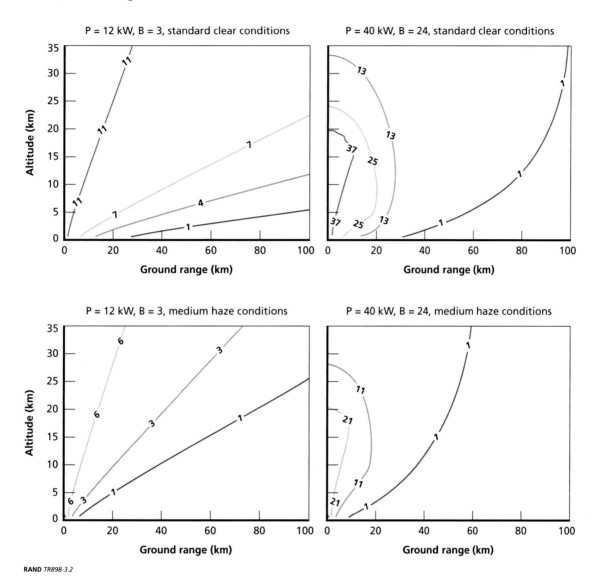

Clouds

Laser beams cannot penetrate an opaque cloud layer. When there are gaps in the cloud cover, the UAV can orbit so as to maintain a cloud-free line of sight to the laser source if its mission permits a choice of orbit. If the beam source is mobile (e.g., on a ship), the beam source may also be able to adjust its position to find cloud gaps. However, if clouds cover the sky at all near-zenith angles over the laser source, the UAV will have to descend below the cloud cover to regain power.

It is difficult to generalize about cloud conditions, since they vary greatly as a function of time and place. Using the PCloudS cloud statistics program, which incorporates several climatological databases (Boehm and Willand, 1995), I looked at cloud cover over two locations,

Baghdad and Taipei. I assumed 80 percent cloud cover as a threshold above which a UAV might have difficulty maintaining a cloud-free line of sight to its laser power source.

Figure 3.3 shows the probability of the sky being less than 80 percent cloudy continuously for 12 hours, from 6:00 pm to 6:00 am, on different nights of the year. In the vicinity of Taipei, there is only a 20 to 30 percent chance that the laser could power the UAV continuously through the night without any serious obstruction from clouds. The winter months in Baghdad are somewhat similar, but in the summer months, the UAV could be deployed with an approximately 80 percent expectation of a continuously clear night. This analysis does not include the possible effects of sandstorms or other obscurants.

If the sky does cloud over but the UAV has significant battery capacity, it might remain above the cloud layer in the hope that conditions will clear before it has to descend. The upper portion of Figure 3.4 shows the probability that 80 percent cloudy conditions will occur and persist for more than 2 hours. The lower portion shows the probability that 80 percent cloudy conditions will occur and persist for more than 6 hours.

A consequence of the findings shown in Figure 3.4 is that with sufficient batteries to endure a 2-hour cloud cover, the UAV would have a 50 percent chance of flying through two consecutive nights (in Iraq in the winter or year-round in Taiwan) without having to descend because of clouds. In Iraq in the summer, the UAV would have a 50 percent chance of flying through two consecutive weeks without having to descend.

If it carried sufficient batteries to endure a 6-hour cloud cover, the UAV would have a 50 percent chance of flying through three to five consecutive nights (or 50 or more consecutive nights in summertime Iraq).

Figure 3.3
Probability That the Sky Will Be at Least 20 Percent Clear for 12 Continuous Hours Between 6:00 pm and 6:00 am

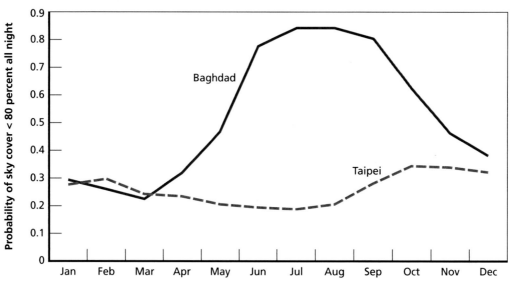

Figure 3.4
Probability of Persistent Cloud Cover Greater Than 80 Percent on a Given Night

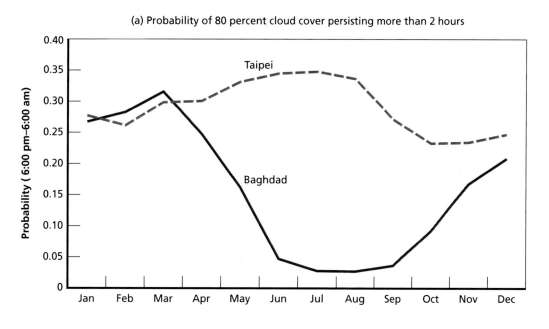

(a) Probability of 80 percent cloud cover persisting more than 2 hours

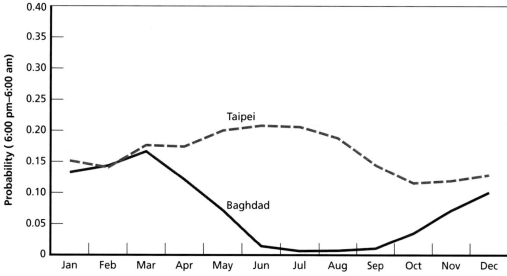

(b) Probability of 80 percent cloud cover persisting more than 6 hours

RAND *TR898-3.4*

If the UAV carried sufficient batteries to endure 12 cloudy hours without laser illumination, it would closely resemble a solar UAV design intended to fly through the night. Reducing the battery pack even by half, to 6-hour endurance, could be quite significant, freeing up to 20 percent of the mass of the aircraft for payload or other systems.

If the UAV's mission were such that it could operate at a range of acceptable altitudes, it could use altitude as a form of energy storage, slowly descending to the minimum acceptable altitude while waiting for the cloud obstruction to clear. An aircraft with the mass and aerodynamic properties of the QinetiQ Zephyr, for example, could take about 2.5 hours to descend from an altitude of 30,000 m to 20,000 m. Of course, the aircraft might still need some batteries to meet electrical needs—for example, to operate the payload—while descending.

One way to circumvent cloud obstruction is to elevate the laser source. Averaged across all times and places on the Earth's surface, the probability of a cloud-free line of sight to a given point on the sky is about 50 percent. As the altitude of the beam source increases, the probability of a cloud-free line of sight increases to 60 percent at 5,000 ft, 70 percent at 20,000 ft, and 95 percent at 40,000 ft (Gregerson, Bangert, and Pappalardi, n.d.). Thus, the cloud problem would be largely negated if the beam source were on an airborne platform at 40,000 ft.

Use of an Airborne Laser Source

The fact that atmospheric beam attenuation and the threat of clouds are much greater at low altitudes promotes the idea of placing the laser itself on an airborne platform. Air-to-air transmission from a source aircraft flying at, say, 40,000 ft (12.2 km) to a UAV potentially flying at much higher altitudes could transmit power across a longer slant range.

Deploying a multikilowatt solid-state laser and the associated optics on an aircraft should be relatively achievable. The Boeing Airborne Laser (ABL) aircraft has a 5,500-kg turret that can house a 1.5-m telescope (Airborne Laser System Program Office, 2003). However, the ABL is a megawatt-class chemical laser requiring adaptive optics to cope with thermal blooming. The sub-megawatt solid-state lasers envisioned here would require simpler optics and would present a less difficult cooling problem. Empirically, commercially available DPSSLs with P kilowatts of output power require about $(0.33 \ P + 0.5)$ m^3/hr of cooling water at 25°C, as shown in Figure 3.5. If the source aircraft were flying at 40,000 ft at Mach 0.5, it would require 17 m^2 of radiator area (with 3-m chord) to keep the cooling water at a temperature of 25°C while removing the waste heat from a 25 percent efficient, 100-kW laser.

Figure 3.5
Cooling-Water Requirements for Commercially Available Lasers

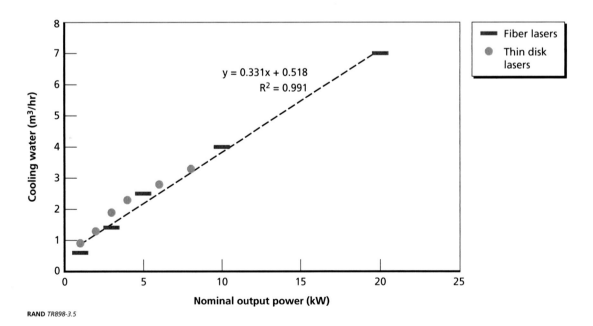

Figure 3.6 shows the flux that could be delivered by a commercially available laser placed on an airborne platform and elevated to 40,000 ft. The y-axis is the altitude of the UAV (which, in principle, can be higher or lower than the altitude of the laser source). Because the beam no longer passes through the lower atmosphere while traveling from the laser to the UAV, the effect of particulate scattering is greatly reduced, and there is no longer much difference between performance on a clear day and performance on a hazy day.

Figure 3.7 shows the flux that could be delivered by a near-state-of-the-art laser aboard an airborne platform. Again, a moderately high-power laser with high beam quality can out-perform a higher-power laser with poorer beam quality. In all the cases studied previously, the limiting factors on laser range were beam quality and atmospheric scattering, but with a high-beam-quality laser at high altitudes, those two factors are minimized. Beam jitter can then

Figure 3.6
Flux Delivered by Commercially Available Fiber Lasers with a 1-m Telescope and 1,060-nm Wavelength from an Airborne Platform at 40,000 ft (kW/m²)

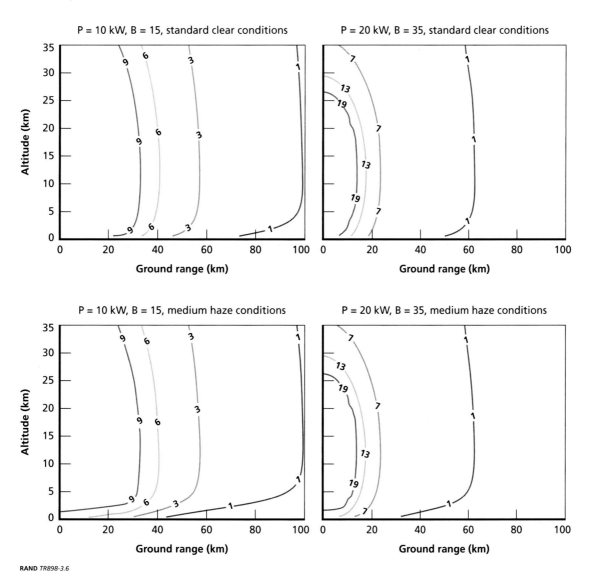

Figure 3.7
Flux Delivered by Current State-of-the-Art Thin Disk Lasers with a 1-m Telescope and 1,060-nm Wavelength from an Airborne Platform at 40,000 ft (kW/m²)

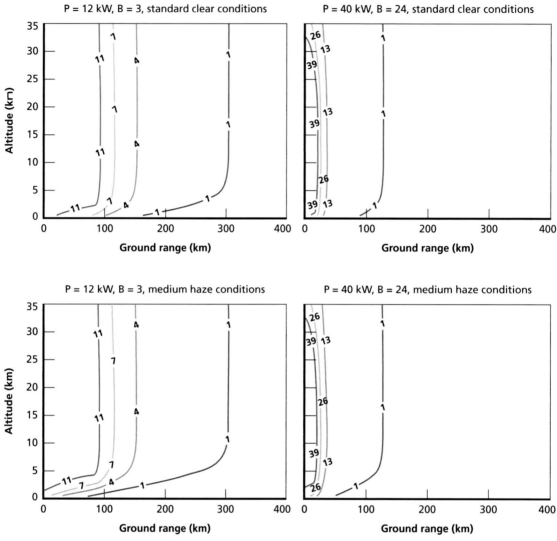

become important, especially since it may be worse from an airborne platform; as explained in Appendix C, for this analysis, I assumed that jitter was 10 μrad from an airborne platform and 6 μrad for a ground-based laser. Nevertheless, a high-quality airborne laser could beam a flux greater than that of sunlight to a UAV hundreds of kilometers away.

This enables entirely different concepts of operation for the UAV. Of course, the persistent availability of the source aircraft would be limited by conventional aircraft endurance. However, the UAV could be supported by multiple source aircraft, potentially hundreds of kilometers apart.

CHAPTER FOUR

Performance Limits of Laser-Powered Aircraft

This chapter discusses some limiting factors to the performance of a laser-PV UAV, including the following:

- Electric motors are heavier than jet engines generating an equivalent amount of thrust, and electric batteries are dramatically heavier than jet fuel with equivalent energy content.
- The amount of power that can be received by the PV cells, and thus ultimately the performance of the UAV, is constrained by thermal operating limits of the PV cells. At extremely high altitudes, laser heating of the PV cells is aggravated by the fact that there is little atmosphere for convective cooling of the UAV.

Thrust/Weight Performance of PV Propulsion

Several studies have examined the power density of electric motors. Brown et al. (2005) analyzed the use of electric motors for aeropropulsion. A high-power-density electric motor was developed for the U.S. Army Tank-Automotive and Armaments Command (TACOM) with a power density of around 3.3 kW/kg. Tesla Motors (2008) also reported building a 52-kg motor with 185 kW of power, or about 3.5 kW/kg, for its electric roadster. Finally, Noth (2008a) observed that many brushless electric motors intended for the high-end hobbyist model airplane market have maximum power densities of around 3.4 kW/kg. (However, Noth also observed that for maximum efficiency, the motors should operate at about 40 percent of maximum power; under those operating conditions, the effective power density is about 1.4 kW/kg.) These power densities compare unfavorably with those of large turbine engines (about 16 kW/kg) but favorably with small-aircraft reciprocating engines (about 0.8 kW/kg).

After reviewing proprietary data for a large number of engines, Brown et al. (2005) concluded that the shaft power of a turbofan at takeoff is typically 0.97 hp/lb of sea-level static thrust, or 1.6 kW/kgf. Furthermore, a purely electric fan would have to deliver 25 percent more shaft power to compensate for the lack of a jet thrust. Thus, about 0.6 kg of high-density electric motor, putting out 2.0 kW of shaft power, is required to generate one kgf of thrust. At laser flux intensities comparable to that of sunlight, about 2 kg of PV cells would be necessary to generate the 2 kW of power. (A smaller mass of PV cells would suffice if the UAV were designed to always operate at higher laser-flux intensities.) So the mass per thrust of the propulsive motor and PV cells together would be about 2.6 kg/kgf.

In comparison, a typical jet engine requires perhaps 0.1 kg of engine to produce 1 kgf of thrust but also burns 0.6 kg of fuel per hour to maintain that thrust. As shown in Figure 4.1,

Figure 4.1
Normalized Weights of Power Source and Energy Storage for Jet Engine and Solar-PV Propulsion

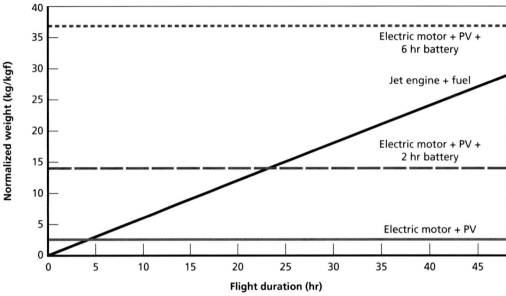

jets have the advantage for short flights, but the pure laser-PV electric motor becomes competitive for flights of more than a few hours' duration, because it needs to burn no fuel. (If the UAV were designed to operate at very high laser intensity and the specific power density of the photovoltaic cells were correspondingly increased, the laser-PV UAV could become competitive even earlier, perhaps on flights of little more than an hour. However, jets will always be the superior technology for very short flights, because of the greater weight of electric motors.)

If the laser-PV aircraft is burdened with batteries to cope with interruptions or variations in beam power, its relative merit is severely damaged, because electric batteries are so heavy. About 5.7 kg of lithium sulfur batteries is required to provide 2 kWh of energy and 1 kgf of thrust for an hour in the absence of laser power. Figure 4.1 presents normalized weights (in kilograms per kilogram-force of thrust) of power source and energy storage for solar-PV UAVs with 2 hours of battery capacity and 6 hours of battery capacity.

Operating Temperature

Since the PV panels lose efficiency at high temperatures, panel heating is potentially the limiting factor on the power that can be provided to the UAV.

In the passive cooling case, the PV panel is on the underside of the wing, but the two sides of the wing are strongly thermally coupled, so the waste heat from the panel can be radiated or convected from both sides. It is possible to imagine an active cooling system in which heat from the panels is pumped to a radiator at a higher temperature than the panel, but that possibility is not considered here.

For simplicity, assume that the surface area of the wing on each side is equal to the area irradiated by the laser. Also for simplicity, assume a nighttime flight when the sun is not present. (If sunlight is present, it can be treated as a source of heat but potentially also as a source of power, much like an increase in the laser flux.)

The adiabatic wall temperature of the air near the panel (i.e., the temperature the air attains without heat transfer from the panel) is given by (Thornton, 1996)

$$T_{aw} = \begin{cases} T_\infty \left[1 + \mathrm{Pr}^{1/2} \left(\dfrac{\gamma-1}{2} \right) M^2 \right] & \text{(laminar flow)} \\ T_\infty \left[1 + \mathrm{Pr}^{1/3} \left(\dfrac{\gamma-1}{2} \right) M^2 \right] & \text{(turbulent flow)} \end{cases} \tag{4.1}$$

where T_∞ is the undisturbed air temperature at the UAV altitude, $\gamma = 1.4$ in the temperature ranges of interest, M is the Mach number of the UAV, and the Prandtl number Pr is the ratio of kinematic viscosity to thermal diffusivity. The Prandtl number of the air as a function of altitude is illustrated in Figure 4.2, based on the standard temperature, density, and viscosity profile of the U.S. Extension to the ICAO Standard Atmosphere, 1958.

The convective cooling of the plate is given by

$$\dot{q}_{\mathrm{conv}} = \rho V c_p \left(T_{aw} - T_{\mathrm{panel}} \right) C_H \tag{4.2}$$

Figure 4.2
Prandtl Number as a Function of Altitude

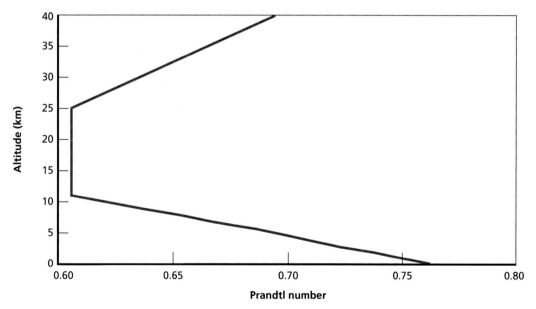

SOURCE: MODTRAN calculations by Kobayashi (2000).
RAND TR898-4.2

where the dimensionless Stanton number C_H is given by

$$C_H = \begin{cases} \dfrac{0.332}{\sqrt{\mathrm{Re}_x^*}}(\mathrm{Pr}^*)^{-2/3} & \text{(laminar flow)} \\[2em] \dfrac{0.185}{\left(\log_{10}\mathrm{Re}_x^*\right)^{2.584}}(\mathrm{Pr}^*)^{-2/3} & \text{(turbulent flow)} \end{cases}$$

$$\mathrm{Re}_x^* = \frac{\rho^* V x}{\mu^*} \tag{4.3}$$

$$\mathrm{Pr}^* = \frac{\mu^* c_p^*}{k^*} \tag{4.4}$$

where ρ^*, c_p^*, and k^* are evaluated at a reference temperature T^*:

$$T^* = T_e[1 + 0.032 M^2 + 0.58(T_w/T_e - 1)] \tag{4.5}$$

and $T_e \approx T_\infty$ is the temperature at the edge of the boundary layer.

Thermal equilibrium is achieved when

$$\phi_{\text{laser}}(1 - \eta_{\text{panel}}) + 2\dot{q}_{\text{conv}} - 2\varepsilon\sigma_{SB}T_{\text{panel}}^4 + \varepsilon\sigma_{SB}T_{\text{sky}}^4 + \varepsilon\sigma_{SB}T_{\text{earth}}^4 = 0$$

where $\sigma_{SB} = 5.67 \times 10^{-8}$ Wm^{-2} K^{-4} is the Stefan-Boltzmann constant. I assume an emissivity ε of 0.9 and an Earth temperature T_{earth} of 288°K. The apparent sky temperature T_{sky} depends on altitude and is lower at higher altitudes, as shown in Figure 4.3.

The PV panel efficiency η_{panel} is itself a function of T_{panel} and also depends on the type of PV material, as discussed in Appendix B. Greater efficiency results in a lower equilibrium temperature, since the panel dissipates less heat. Figure 4.4 shows the resulting equilibrium temperatures for a laser flux of 5,000 W/m^2 for both CIS PV cells and high-efficiency custom InGaAs cells. (The flux value of 5,000 W/m^2 is chosen only as an example.) Over most of the range, increasing Mach number lowers the equilibrium temperature via increased convective cooling. The lowest equilibrium temperatures are achieved at an altitude of around 10 km. At lower altitudes, the air and apparent sky temperatures are higher; at higher altitudes, the thinner atmosphere does a poorer job of convective cooling.

By considering different values for the laser flux ϕ, we can find an optimal flux intensity to deliver the maximum power to the UAV. That is, at a given altitude and Mach number, and for a given choice of PV converter, there is a value of ϕ beyond which more laser intensity would be counterproductive—it would heat up the UAV too much and would result in less usable power.

While a greater maximum power can theoretically be delivered to the UAV at higher Mach numbers (because of convective cooling), the UAV would also require greater power to overcome increased drag, so more power would not necessarily be available for use by the payload.

Figure 4.3
Effective Sky Temperature as a Function of Altitude

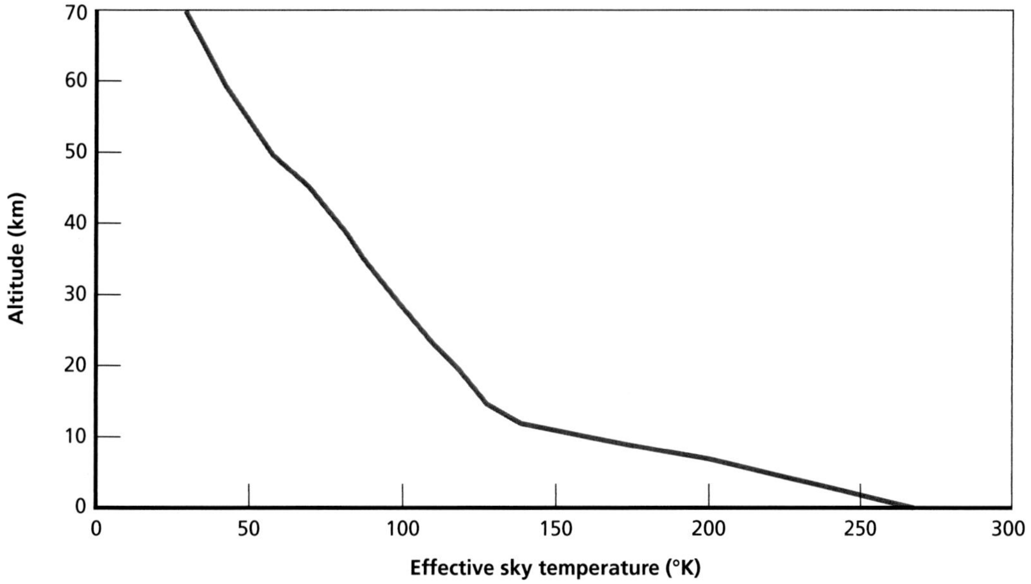

SOURCE: MODTRAN calculations by Kobayashi (2000).
RAND TR898-4.3

Figure 4.5 shows the theoretical maximum useful laser flux irradiating a UAV with CIS PV panels, along with the resulting optimal operating temperature. Figure 4.6 shows the resulting maximum usable power derivable from CIS PV panels as function of altitude and Mach number. Figure 4.7 and Figure 4.8 show the equivalent plots for a UAV using customized InGaAs PVs.

The best existing solar panels can produce almost 400 W/m^2 under optimal conditions, but the optimal arrival angle would almost never be realized for a solar aircraft; perhaps 200 to 300 W/m^2 would be a more realistic goal for usable power collected from solar panels. We see that when using CIS PVs, the laser-powered UAV is theoretically capable of receiving around twice as much power per unit area as a solar-powered UAV. If customized InGaAs PVs are used, the laser-powered UAV can achieve perhaps 10 times more power per unit area than a solar-powered UAV.

The Effect of Increased Power on Aircraft Performance

Power delivered by a laser that exceeds the power available from sunlight will, of course, expand the possible performance envelope of a PV aircraft. To quantify the increased performance, it is necessary to consider how the propulsion, battery, and PV subsystems will be affected by an increase in power.

The lift and drag forces on the UAV are

$$L = C_L \frac{\rho}{2} S V^2$$

$$D = C_D \frac{\rho}{2} S V^2$$

(4.6)

Figure 4.4
Equilibrium Temperature as a Function of Mach Number and Altitude for a Laser Flux of 5,000 W/m² (°K)

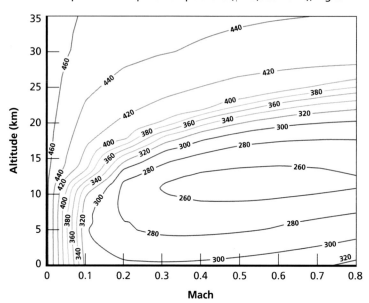

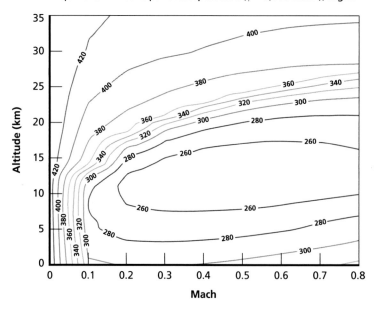

Figure 4.5
Maximum Useful Laser Flux and Resulting Equilibrium Temperature for CIS PVs

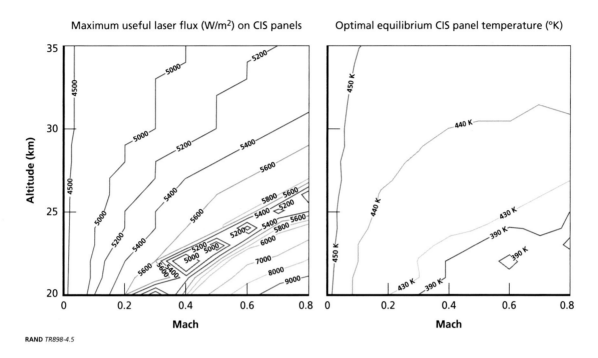

Figure 4.6
Maximum Usable Power Delivery to CIS PVs

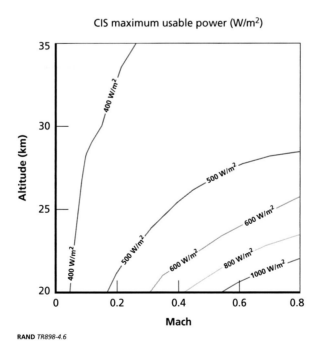

Figure 4.7
Maximum Useful Laser Flux and Resulting Equilibrium Temperature
for InGaAs PVs

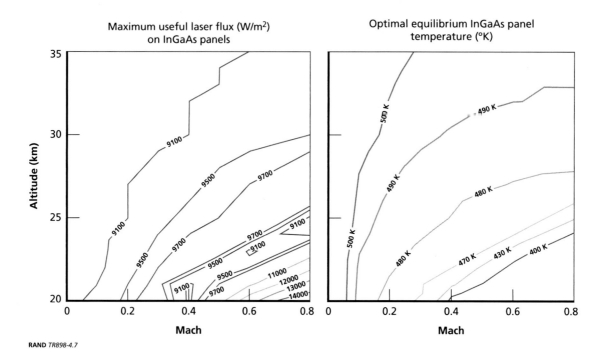

RAND *TR898-4.7*

Figure 4.8
Maximum Usable Power Delivery to InGaAs PVs

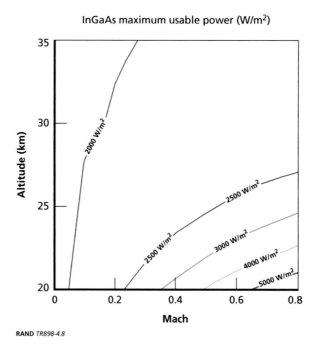

RAND *TR898-4.8*

where C_L and C_D are the coefficients of lift and drag,[1] ρ is the air density, S is the wing area, and V is the true airspeed. In level flight, the lift L equals the weight of the aircraft mg, and the power expended by the propulsion system equals VD/η_{prop}:

$$P_{prop} = VD / \eta_{prop} \tag{4.7}$$

The coefficient of lift C_L can be controlled, within limits (e.g., by adjusting the angle of attack), and C_D varies with C_L:

$$C_D = C_{D0} + \frac{C_L^2}{e\pi(AR)} \tag{4.8}$$

Following Noth (2008a), assume that C_{D0} equals 0.0191 and that the "Oswald efficiency factor" e is equal to 0.9 for PV aircraft. Assume also that in normal flight, C_L cannot exceed 1.5.

The aircraft's operating envelope is therefore defined by (1) the stall limit, where $C_L = 1.5$ and where the aircraft cannot go any slower without losing altitude, and (2) the power limit, where the aircraft cannot go any faster with the power available against the drag induced by the necessary lift. The laser power source can affect the power limit both directly (by increasing the power available) and indirectly (e.g., by reducing the need to carry batteries, which reduces the necessary lift and thus the induced drag).

Noth (2008a) derived scaling equations for the airframe and various subsystems of a solar-powered aircraft. For gossamer aircraft like the Helios and Zephyr, he suggested the following empirical equation for the mass of the airframe, $m_{airframe}$:

$$\begin{aligned} m_{airframe} &= C_{gossamer} b^{3.1} AR^{-0.25} \\ &= C_{gossamer} S^{1.55} AR^{1.3} \end{aligned} \tag{4.9}$$

where S is the wing area, b is the wingspan, AR is the aspect ratio, and the constant $C_{gossamer} = 0.00224$ kg/m$^{3.1}$.

Again following Noth, assume that the mass of the electric motors is 0.70 kg/kW of nominal power,[2] the mass of the gearbox is 0.20 kg/kW, the mass of the propeller is 0.25 kg/kW, and the mass of the motor controller is 0.06 kg/kW, for an overall propulsion subsystem mass of 1.21 kg/kW. Also assume that the propeller is 85 percent efficient, the motor is 85 percent efficient, the gearbox is 97 percent efficient, and the controller is 95 percent efficient, for an overall propulsion efficiency η_{prop} of 66.6 percent. Assume that the mass of the PV cells, including covering, is 0.58 kg/m^2, and that the energy density of the electric storage (say, lithium sulfide batteries) is 350 Wh/kg.

---

[1] This calculation does not take any explicit account of drag on non-wing elements that is not proportional to wing area. One could consider any such drag to be subsumed in the minimum drag coefficient C_{D0}.

[2] The implicit assumption here is that the nominal power expended during level flight is about 40 percent of the electric motor's maximum power.

The operating envelopes for several possible PV aircraft are plotted in Figures 4.9 through 4.13. Table 4.1 summarizes the different assumptions behind the operating envelopes.

Figure 4.9
Nighttime Operating Envelope of Zephyr-Like Solar Aircraft

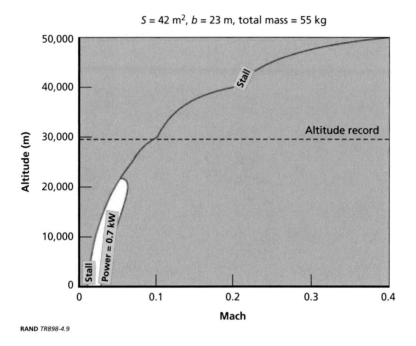

RAND *TR898-4.9*

Figure 4.10
Nighttime Operating Envelope of Zephyr-Like Aircraft
with a Laser Providing 700 W of Power

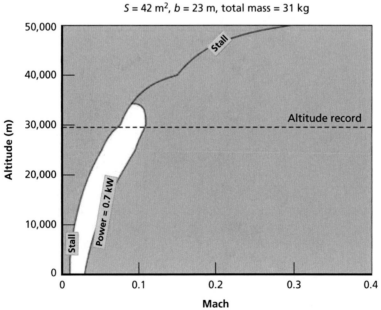

RAND *TR898-4.10*

Figure 4.11
Nighttime Operating Envelope of Zephyr-Like Aircraft
with a Laser Providing 2.8 kW of Power

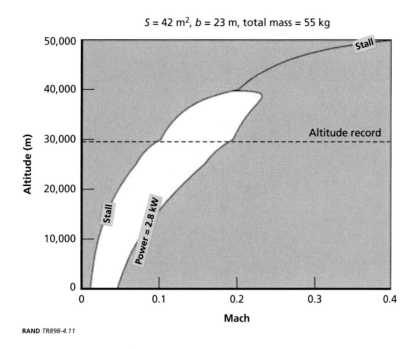

$S = 42\ m^2,\ b = 23\ m,\ total\ mass = 55\ kg$

Figure 4.12
Nighttime Operating Envelope of Zephyr-Like Aircraft
with a Laser Providing 16 kW of Power

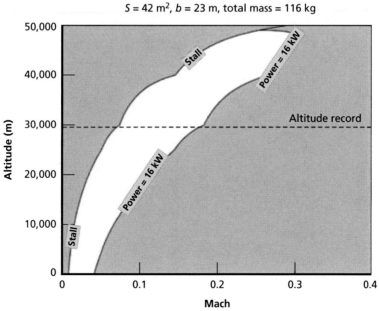

$S = 42\ m^2,\ b = 23\ m,\ total\ mass = 116\ kg$

Figure 4.13
Nighttime Operating Envelope of Non-Gossamer Aircraft
with a Laser Providing 22 kW of Power

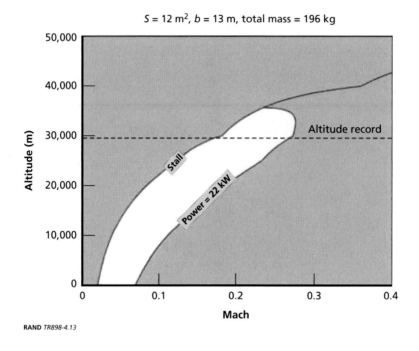

Table 4.1
Parameters for Some PV Aircraft Concepts

Concept	Solar Only	Laser-Enabled			
	Solar UAV similar to Zephyr (Figure 4.9)	Laser replaces sunlight (Figure 4.10)	4x power from laser; 4x payload mass (Figure 4.11)	Power near thermal limit of CIS (Figure 4.12)	Power near thermal limit of InGaAs (Figure 4.13)
Wing area S, m^2	41.85	41.85	41.85	41.85	11.6
Wingspan b, m	23	23	23	23	12.6
Construction	Gossamer	Gossamer	Gossamer	Gossamer	Non-gossamer
Nominal propulsive power, W	700	700	2,800	16,000	22,000
PV usable power density, W/m^2	50 (average)	100	200	400	2,000
Airframe mass, kg	19.8	19.8	19.8	19.8	60.3
Propulsion mass, kg	0.8	0.8	3.4	19.4	26.6
PV mass, kg (area, m^2)	8.1 (14.0)	4.1 (7.1)	8.1 (14.0)	23.2 (40.0)	6.4 (11.0)
Battery mass, kg (duration, hr)	24 (12)	4 (2)	16 (2)	45.7 (1)	62.9 (1)
Payload mass, kg	2	2	8	8	40
Total mass, kg	54.7	30.7	55.3	116.1	196.1
Max. altitude, m	21,700	34,300	39,700	49,000	35,600

Figure 4.9 illustrates the nighttime operating envelope of an indefinite-endurance solar-powered aircraft, intended to resemble the larger-model QinetiQ Zephyr of 2010, with a total mass of 55 kg, of which 2 kg is payload and 24 kg is battery. The nominal operating power of the propulsion system in level flight is 700 W, which the batteries can sustain for 12 hours. Under these conditions the highest achievable altitude is 21,700 m.

Figure 4.10 illustrates the nighttime operating envelope of a similar aircraft with solar power replaced by an equivalent laser power source. Since the laser is presumed to be available throughout the night, there is no need for a large battery capacity; here, the required battery duration is assumed to be 2 hours instead of 12 hours. Also, even if the laser supplies only the same usable power per square meter as sunlight, the laser-powered aircraft requires fewer square meters of PV cells than the solar-powered aircraft, because the laser-powered aircraft does not need excess power during the day to charge batteries for the night. With the mass savings from reduced battery and PV capacity, this aircraft can sustain flight indefinitely at altitudes up to 34,300 m. (The same or similar altitudes should be achievable by a solar-powered aircraft intended to fly only during the day, such as the Helios HP01, whose altitude record is indicated in the figure.)

Now, suppose that instead of merely replacing the power source of the solar Zephyr, the laser supplies a power budget four times larger (2,800 W) with a usable array power density of 200 W/m². This seems comfortably achievable given what has been discussed of available laser power, atmospheric transmission, and PV conversion efficiency. Figure 4.11 illustrates the operating envelope of the aircraft, with the propulsion system resized to take advantage of the larger power budget and the payload quadrupled to 8 kg. This version of the aircraft has total mass similar to that of the solar Zephyr, but most of the batteries have been replaced by payload and propulsion. This higher-powered aircraft can sustain altitudes of 39,700 m, significantly above existing altitude records.

Finally, consider some extreme designs approaching the thermal limits described in Figures 4.5 through 4.8. Figure 4.12 shows the operating envelope for a Zephyr-like aircraft with a laser supplying 16 kW of power—about 400 W/m² across the whole wing area, which is close to the thermal limit for CIS PVs. At this high power level, the weight of even a short-duration battery capacity becomes burdensome, so the requirement for battery duration is assumed further reduced to 1 hour. This gossamer aircraft could carry an 8-kg payload at an altitude of 49,000 m.

Figure 4.13 shows the performance of an aircraft that is *not* a large gossamer airframe like Helios or Zephyr. Instead, the mass of this airframe is assumed to match the empirical curve from Noth (2008a) that fits the "best 5 percent" of sailplane designs:

$$m_{\text{airframe}} = C_{\text{sailplane}} b^{3.1} AR^{-0.25}$$
$$= C_{\text{sailplane}} S^{1.55} AR^{1.3} \tag{4.10}$$

where the constant $C_{\text{sailplane}} = 0.0449$ kg/m$^{3.1}$. Thus, this aircraft is about 20 times heavier than a gossamer aircraft of similar dimensions, although by the standards of conventional aircraft, it would still be considered very sophisticated, lightweight construction. The smaller wing area of this aircraft is assumed to receive 2,000 W/m² of usable laser power, which is close to the thermal limit for tailored InGaAs PV cells. The payload is also assumed to increase to

40 kg. The battery capacity is assumed to be 1 hour. With 22 kW of power, this aircraft can sustain an altitude of up to 35,600 m, still above existing altitude records.

Conclusion

Within the limits imposed by available laser technology and the thermal properties of PV cells, laser-powered aircraft should be able to carry greater payloads than solar-powered aircraft, while sustaining substantially higher altitudes than any existing winged aircraft, solar-powered or otherwise. At higher power densities, a laser-powered aircraft could also afford a heavier, more conventionally constructed airframe, which might be more robust than the extraordinarily light gossamer airframes needed for large solar aircraft.

The various configurations suggested in Table 4.1 are only examples. A designer might choose to sacrifice maximum altitude for a heavier payload or might divert much of the power from the laser for use by the payload rather than by the vehicle. However, the sustained high altitudes illustrated in Figures 4.10 through 4.13 are particularly compelling, since they are not easily duplicated by competing concepts.

CHAPTER FIVE

Conclusions and Future Directions

Conclusions

It is feasible with existing technology to build a laser-powered UAV with performance characteristics beyond the envelope of current aircraft. Specifically, a laser-powered UAV could plausibly fly at higher altitudes than any existing winged aircraft, while also carrying a larger payload than can be carried by a solar-powered UAV. Even with conservative assumptions and commercial off-the-shelf lasers and PV cells, a laser-powered UAV could carry four times the payload and achieve 80 percent greater nighttime altitude than a solar-powered UAV of the same size and total mass. Even bigger performance gains, e.g., maximum altitudes approaching 50 km, could be achievable with more aggressive assumptions.

The most suitable type of laser for this application would be either a fiber laser or a thin disk DPSSL. Simple diode lasers, such as those used in a NASA proof-of-concept experiment, are not bright enough to scale up to a UAV at altitudes of practical interest. Megawatt-class chemical lasers, on the other hand, are unnecessarily powerful and also emit too far into the infrared; they would introduce operating difficulties without providing any real benefit to the UAV, because of thermal limits on the PV cells.

CIS PV cells could be used for a strictly off-the-shelf design. However, a better option would be custom-made InGaAs PV cells tailored to the wavelength of the laser power source.

Assuming state-of-the-art PVs and passive cooling, the usable power delivered to the UAV platform is limited by thermal considerations to between 2,000 and 3,000 W/m², about 10 times the useful power available to a solar-powered aircraft. This thermal limit could perhaps be circumvented by an active cooling scheme, which is not considered here.

The persistence of the platform would be indefinite but limited by cloudy weather (in the case of a ground-based power source) or by the availability of an airborne laser power source. If the laser were beamed from the ground or from a ship, the UAV would be closely "tethered" to the beam source and (to receive useful amounts of power) would have to fly in an orbit within a few tens of kilometers of it. Also, clouds might interrupt the beam and force the UAV to descend below the cloud layer.

Both of these problems could be circumvented by placing the laser on a conventional aircraft, so that the UAV would be powered by an air-to-air transmission. In this case, the "tether" from the UAV to the power source could be much longer (hundreds of kilometers) and clouds would no longer be a likely threat. Deploying the laser source on an aircraft should be technologically feasible, though, of course, flying the source aircraft imposes an additional operational burden. The persistence of the UAV would then become tied to the persistence of the source aircraft.

Because of the weight of electric motors, PV cells, and electric batteries, jet propulsion is generally a superior technology except for missions requiring extreme endurance or extremely high altitude. Because of the effort required to support the laser-PV UAV above the cloud layer, the "ultra-long persistence" argument is not very compelling. However, the laser-PV concept could be worth further consideration if an important mission were identified for an air vehicle with ultra-high operating altitude and reasonable persistence and payload. Some possible applications include ultra-high-altitude observation stations or communication relays and flocks of high-altitude sensor probes powered remotely from a large aircraft "mother ship."

Future Directions

This report has focused on the physical parameters of flight—altitude, range, persistence, and power—that are possible for a laser-PV aircraft with current technology. Some questions that could be explored by a future study include:

- Are there specific missions or concepts of operation where persistence at extreme altitude makes the laser-PV concept attractive?
- What specific payloads could be supported within the power and weight constraints of a laser-PV aircraft?
- What would be the costs of building and operating a laser-PV aircraft?
- Are there any mission concepts in which only "fair weather" performance is required? In other cases, would it be acceptable to mitigate vulnerability to cloudy weather through the use of batteries? Or must all-weather operations be provided through the use of an airborne laser source?
- What should be the duty cycle of the laser source? Could one source serve multiple UAVs?
- In military applications, could a laser-PV UAV have any degree of stealth, or would it necessarily have a large infrared and/or radar signature? Would the laser beam also reveal the location of the laser source? If the UAV cannot be made stealthy, could it be made semi-expendable?

Answers to these questions can help to determine whether the performance niche opened by this concept is operationally valuable and worth pursuing or is merely a curious technology "stunt" waiting to happen.

Available Types of Laser

Semiconductor lasers or laser diodes are compact, efficient, mass-produced, and inexpensive but are usually considered to have low power. Diode lasers have the advantage of being able to generate light at wavelengths around 800 nm, which is ideal for the highest efficiency conversion by PV cells, as shown in Figure 1.8. Unfortunately, as shown in Table 2.1, laser diode arrays are not bright enough to achieve the minimum threshold $A_{source} R_{source} > 1.6 \times 10^{12}$ W/sr. More precisely, compensating for the low radiance of the laser diodes would require about 100 m^2 of telescope aperture, somewhat larger than that of the world's largest telescopes—and this would be to achieve only the most minimally interesting performance by a laser-powered UAV.

Spectral-beam combining is a scheme in which the output of laser diodes is combined to achieve higher radiance. The radiance from a laser of this type is just marginally adequate to meet the threshold requirements of a laser-powered UAV, using a 1-m telescope.

More promising are lasers in which semiconductor diodes are used to pump a lasing medium (e.g., rare Earth minerals such as neodymium, erbium, ytterbium, or holmium) suspended in a crystalline solid. Since heat dissipation from the solid is an important problem, it is helpful to shape the material into a form with a high ratio of surface area to volume. Two types of DPSSLs are fiber lasers, in which the lasing material has been stretched into a long, thin fiber, and thin disk lasers, in which the material forms a thin slab.

DPSSLs have the disadvantage that they produce light at wavelengths greater than 1 μm. As shown in Figure 1.8, this is too long for efficient conversion by silicon or GaAs PV cells. However, other PV materials such as germanium or CIS have a spectral response that extends to the near-infrared and can be used to convert the beam to electricity, though perhaps at some penalty in efficiency.

Fiber lasers with up to 50 kW of power and efficiency greater than 25 percent are available commercially and are in widespread industrial use for cutting and welding (IPG, 2008). Beam quality is good for single fibers with 2 kW of power and not as good for higher-power fiber lasers; however, commercially available fiber lasers appear adequately powerful and bright for a laser-PV UAV.

Thin disk lasers with 8 kW of power, like the one shown in Figure 2.2, are commercially available (TRUMPF, 2008). Numerical models show that power output of more than 40 kW from a single disk is possible (Giesen and Speiser, 2007). Thin disk lasers can achieve very high beam quality and also appear potentially adequate for a laser-PV UAV. Thin disk lasers may be easier than fiber lasers to scale to 100 kW or more while maintaining good beam quality. In June 2008, Boeing fired a thin disk laser with a beam quality "suitable for a tactical weapon system" at more than 25 kW for multisecond durations (Selinger, 2008).

The most powerful lasers produced to date are chemical lasers such as the chemical oxygen iodine laser (COIL) and deuterium fluoride or hydrogen fluoride lasers that use flowing chemicals as the laser medium. These have been scaled up to megawatt power levels but all lase in the mid- to far-infrared, so the efficiency of PV conversion would be penalized.

One more laser technology that could be advantageous in the future is the diode-pumped alkali laser (DPAL). DPALs use an alkali gas instead of a solid-state lasing medium. The alkali gas is mixed with a buffer gas of high-pressure helium which helps conduct the heat to a cooling surface. The gaseous medium can produce very high beam quality because of the low variance of index of refraction with temperature. The gaseous DPALs are a promising future choice for laser-power beaming, since they could be scaled to 100-kW levels with high brightness at wavelengths consistent with peak conversion by silicon or GaAs PV cells (Krupke et al., 2003). DPALs could even be scaled to megawatt power levels if the alkali gas were continuously circulated through the lasing cell rather than remaining static (Krupke et al., 2004). Such high power levels have not been realized yet, however. In 2008, researchers at the U.S. Air Force Academy produced a diode-pumped cesium laser with 48 W of output power (Zhdanov et al., 2008).

Researchers at Lawrence Livermore National Laboratory (Page et al., 2005) estimate that a high-power DPAL could have a weight-to-power ratio on the order of 7 kg/kW. This would be a substantial improvement over the weight-to-power ratio of existing commercial fiber lasers, as illustrated in Figure A.1.

Figure A.1
Weight as a Function of Power for Commercially Available IPG YLR-Series Fiber Lasers and Projected Future DPALs

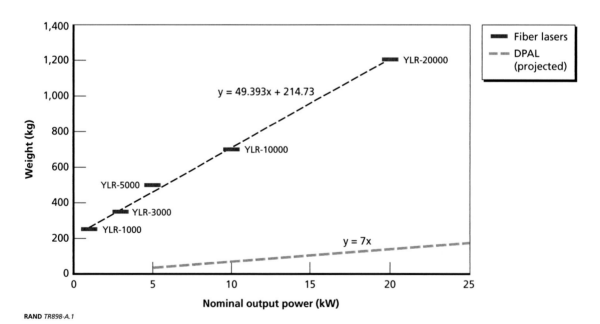

Photovoltaic Conversion of Laser Light

Photovoltaic cells can be more efficient under light at a well-matched frequency than they are under sunlight. Landis (1993) offers the following conversion[1] from solar spectrum efficiency to efficiency under monochromatic illumination:

$$\eta_{\text{laser}} \approx \eta_{\text{solar}} \frac{P_{\text{sun}}}{J_{\text{sc}}} (QE) \frac{\lambda}{1240\ \text{nm}} \left[1 + \frac{25\ \text{mV}}{V_{\text{oc}}} \ln \left(QE \frac{\lambda}{1240\ \text{nm}} \frac{\phi}{J_{\text{sc}}} \right) \right] \tag{B.1}$$

where η_{laser} is the efficiency of the PV cell under laser illumination, η_{solar} is the efficiency under solar illumination, P_{sun} is the solar intensity under which η_{solar} was measured, J_{sc} is the short-circuit current density and V_{oc} is the open-circuit voltage under those solar conditions, ϕ is the laser intensity, and QE is the internal quantum efficiency at the laser wavelength λ. The PV efficiency improves somewhat with increased intensity. Equation B.1 is valid for intensities up to some limit where series resistance becomes a significant factor in efficiency.[2] In Table B.1, this equation is applied to estimate the monochromatic efficiency for three PV materials.

The most developed and most widely produced PVs are GaAs and silicon. These cells can have greater than 50 percent efficiency under light of the right wavelength. Unfortunately, these materials are best suited to wavelengths around 850 nm and are not responsive to the wavelengths of more than 1 μm that are characteristic of current high-energy lasers.

Thin-film CIS cells with η_{solar} = 12 percent and projected η_{laser} = 17 percent have the advantages of being in large-scale production and the ability to be deposited on thin titanium foils (Powalla, Dimmler, and Groß, 2005). Such thin-foil PVs have an areal density of 0.2 kg/m^2 (Woods, Ribelin, and Armstrong, 2007) and are flexible, thus lending themselves to application on the aerodynamic surfaces of a UAV.

InGaAs cells can also be used to convert laser light at longer wavelengths. The energy bandgap of InGaAs cells can be varied by adjusting the relative fractions of indium and gallium (Aldair et al., 1996). Light at λ = 1,550 nm can be converted with an efficiency of 34 percent, while light at λ = 2,100 nm can be converted at 22 percent efficiency (Wojtczuk, 1997). Officials at Spectrolab state that they could design an InGaAs device around the λ = 1,060-nm

---

[1] Landis explains the derivation of his equation but has a misprint in the result, which is corrected here.

[2] A PV cell can be modeled as a current source in parallel with a diode and a shunt resistance R_{SH} and in series with a series resistance R_s. Since the power lost to the series resistance is quadratic with the current through the system, the effect of the series resistance is most significant at high intensities and high currents.

wavelength with an efficiency of about 50 percent. Some implementation costs would be incurred, since the device is not currently in production.[3]

Table B.1 shows estimated monochromatic efficiencies η_{laser} for three PV materials: GaAs (Landis, 1993), germanium (Posthuma et al., 2007), and CIS (Olsen et al., 1997). Germanium cells are projected to have an efficiency of only about 13 percent at λ = 1,060 nm. However, Nagashima et al. (2005) developed a germanium cell with a projected efficiency of 25 percent at λ = 1,500 nm.

The efficiencies given so far are for a nominal operating temperature of $25\,°C$ ($298\,°K$). The efficiency of PVs drops approximately linearly with increasing temperature:

$$\eta = \eta_{nominal} + \frac{d\eta}{dT}(T - T_{nominal}) \tag{B.2}$$

Furthermore, the magnitude of the (negative) normalized temperature coefficient $(1/\eta)d\eta/dT$ is proportional to the bandgap of the material, i.e., the effect of temperature is worse for PVs tuned to longer wavelengths. For a given cell, the temperature coefficient under monochromatic light should be roughly the same as that under solar light, although not identical (Landis, 1993).

The normalized temperature coefficient for commercial CIS thin films under solar radiation is in the range of −0.33 percent/$°K$ to −0.41 percent/$°K$ (Mohring et al., 2004). The normalized temperature coefficient for Spectrolab's advanced triple junction cells is about −0.2 percent/$°K$, varying slightly as a function of input intensity. Since the coefficient scales linearly with the optimal laser wavelength for the material, it is reasonable to suppose that an InGaAS device tailored to the 1,060-nm wavelength would have a normalized temperature coefficient of between −0.25 percent/$°K$ and −0.3 percent/$°K$.

The most efficient cells for solar conversion are multijunction cells with three junctions on top of each other, each one absorbing photons from a different part of the solar spectrum. These will not work under monochromatic light of a single wavelength, because when only one of the three junctions is illuminated, the other two are in the high-impedance regime and

Table B.1
Estimated Efficiencies Under Laser Illumination of Three PV Materials

PV Material	Estimated Efficiency							
	J_{sc} (solar), mA/cm^2	V_{oc} (solar), V	P_{sun}, W/m^2	η_{solar}, percent	Suitable Laser Wavelength λ, nm	Quantum Efficiency at λ	η_{laser} at ϕ = 1,000 W/m^2, percent	η_{laser} at ϕ = 10,000 W/m^2, percent
Gallium arsenide	33.1	1.033	1,370	21.7	850	0.85	53.1	56.1
Germanium	46.4	0.269	1,000	7.8	1,060	0.85	12.7	15.4
CIS	40.7	0.439	1,000	12.0	1,060	0.68	17.5	19.7

---

[3] Dmitri Krut (Spectrolab), personal communication, 2008.

block the current flow. Unfortunately, this would make it difficult to use these cells for both solar conversion and laser conversion.[4]

---

[4] Of course, the aircraft could possibly have advanced triple junction cells for solar collection on its top surface and single junction cells on its underside, to be sustained by sunlight during the day and laser power at night. Although there would be a weight penalty for the "redundant" PV cells, it would be significantly less than the weight of an extended battery-storage capacity, as described in Chapter Four.

Laser-Beam Transmission Equations

The laser flux arriving at the target is

$$\phi = 0.86 \frac{4P\exp(-\varepsilon L)}{\pi L^2 (\sigma^2_{\text{diffraction}} + \sigma^2_{\text{turbulence}} + \sigma^2_{\text{jitter}} + \sigma^2_{\text{bloom}})}$$

$$= \frac{3.44 P\exp(-\varepsilon L)}{\pi L^2 (\sigma^2_{\text{diffraction}} + \sigma^2_{\text{turbulence}} + \sigma^2_{\text{jitter}} + \sigma^2_{\text{bloom}})} \tag{C.1}$$

where P is the laser power, L is the distance to the target, ε is the atmospheric extinction of the beam due to absorption or scattering, and $\sigma_{\text{diffraction}}$, $\sigma_{\text{turbulence}}$, σ_{jitter}, σ_{bloom} are the beam-spreading factors for diffraction, turbulence, beam jitter, and thermal blooming, respectively (Friedman and Miller, 2003; Kare, 2004).

The laser is assumed to have an approximately Gaussian profile with 86 percent of the beam power within the angular diameter:

$$\Delta\theta_{\text{full}} = \sqrt{\sigma^2_{\text{diffraction}} + \sigma^2_{\text{turbulence}} + \sigma^2_{\text{jitter}} + \sigma^2_{\text{bloom}}} \tag{C.2}$$

Diffraction

The amount of beam spreading due to diffraction is

$$\sigma^2_{\text{diffraction}} \simeq \left(\frac{2}{\pi}\frac{B\lambda}{w_0}\right)^2$$

$$\simeq \left(\frac{2B\lambda}{D}\right)^2 \tag{C.3}$$

where w_0 is the diameter of the waist (smallest) size of the beam in the beam-forming optics, and $D = \pi w_0$. (If the aperture has diameter $D = \pi w_0$, 99 percent of the power in the beam is transmitted through the aperture; Friedman and Miller, 2003.)

Of the available laser systems listed in Table 2.1, those of particular interest are the multikilowatt thin disk and fiber lasers with wavelength $\lambda = 1,060$ nm and beam quality between

B = 3 and B = 35. Assuming D = 1 m, for these lasers, $\sigma_{\text{diffraction}}$ ranges from 6.4 μrad (B = 3) to 74 μrad (B = 35).

Absorption and Scattering

Attenuation of the laser beam is caused by both absorption (primarily by atmospheric molecules) and scattering (also by molecules, but primarily by larger suspended particles such as water, dust, or smoke):

$$\varepsilon = \alpha_{\text{absorption}} + \alpha_{\text{scattering}} \qquad (C.4)$$

Scattering generally dominates at the near-infrared wavelengths of interest. Cook (2004) estimated the absorption coefficient at 1,060-nm wavelength to be less than 0.0004 km^{-1} across a wide range of maritime conditions. Atmospheric attenuation coefficients at sea level are shown for certain wavelengths and visibility conditions in Table C.1 (Burle Industries, 1974). Scattering drops off exponentially with altitude with a scale height of about 1,200 m (Burle Industries, 1974), as shown in Figure C.1.

Turbulence

The index-of-refraction structure constant C_n^2 is a measure of variations in the index of refraction caused by small-scale temperature variations in the atmosphere. The Hufnagel-Valley 5/7 model estimates C_n^2 as a function of altitude h in meters as follows:

$$C_n^2 = 8.2 \times 10^{-26} \left(\frac{h}{1000}\right)^{10} W^2 e^{-h/1000} + 2.7 \times 10^{-16} e^{-h/1500} + 1.7 \times 10^{-14} e^{-h/100} \qquad (C.5)$$

where W, the wind correlating factor, is set at 21 (Friedman and Miller, 2003). For nearly horizontal paths, along which C_n^2 is nearly constant, we find the transverse coherence distance (Friedman and Miller, 2003):

$$\rho_{\text{coherence}} = \left(\frac{\lambda^2}{4 C_n^2 L}\right)^{3/5} \qquad (C.6)$$

Table C.1
Sea-Level Attenuation Coefficients ε for Key Laser Wavelengths (km⁻¹)

Wavelength, nm	Attenuation Coefficient, ε (km⁻¹)		
	Standard Clear	Medium Haze	Heavy Haze
850	0.13	0.58	0.96
1,060	0.116	0.53	0.88

Figure C.1
Attenuation Coefficient ε as a Function of Altitude for Key Wavelengths

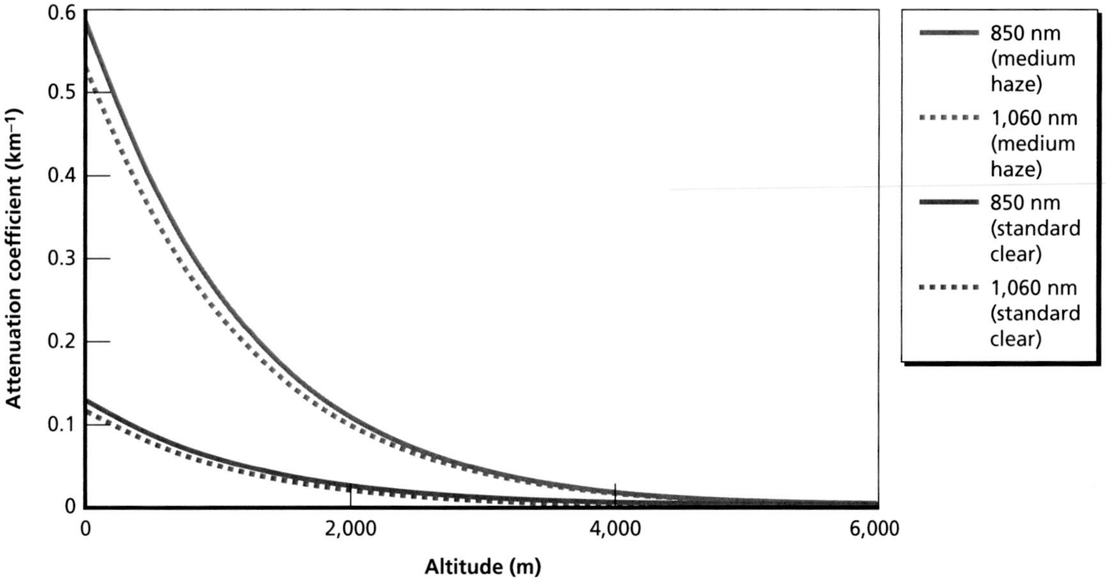

RAND *TR898-C.1*

and then estimate the beam spreading caused by turbulence:

$$\sigma^2_{\text{turbulence}} = \frac{\lambda^2}{4\pi^2 \rho^2_{\text{coherence}}} \tag{C.7}$$

For paths that are not horizontal and along which C_n^2 is not constant, we compute the Fried parameter r_0 as follows:

$$r_0 = \left[0.423 k^2 \sec^2 \beta \int_0^L C_n^2(h) dl \right]^{-3/5} \tag{C.8}$$

where $k = 2\pi/\lambda$ is the propagation constant, β is the zenith angle, and L is the path length.

The amount of beam spreading caused by atmospheric turbulence can be approximated as (Tyson and Ulrich, 1993)

$$\sigma^2_{\text{turbulence}} = \begin{cases} 0.182(\sigma_{\text{diffraction}} / B)^2 (w_0 r_0)^2 & \text{if}(w_0 / r_0 < 3.0 \\ (\sigma_{\text{diffraction}} / B)^2 \left[(w_0 / r_0)^2 - 1.18(w_0 r_0)^{5/3} \right] \text{if}(w_0 / r_0) \ge 3.0 \end{cases}$$

This second expression for $\sigma_{\text{turbulence}}$ does not apply to horizontal paths; it blows up because of the secant of zenith angle. I use the second approximation only when $\sec^2(\beta) < 1.5$ (paths within 35° of zenith).

Figure C.2 shows the resulting values of $\sigma_{\text{turbulence}}$ for a ground-based laser with $\lambda = 1,060$ nm and a 1-m telescope. I did not attempt to smooth or reconcile the two approximations for $\sigma_{\text{turbulence}}$, and there appears to be a minimum at the "seam" between the near-zenith

Figure C.2
Beam Divergence Caused by Turbulence for a Ground-Based Laser

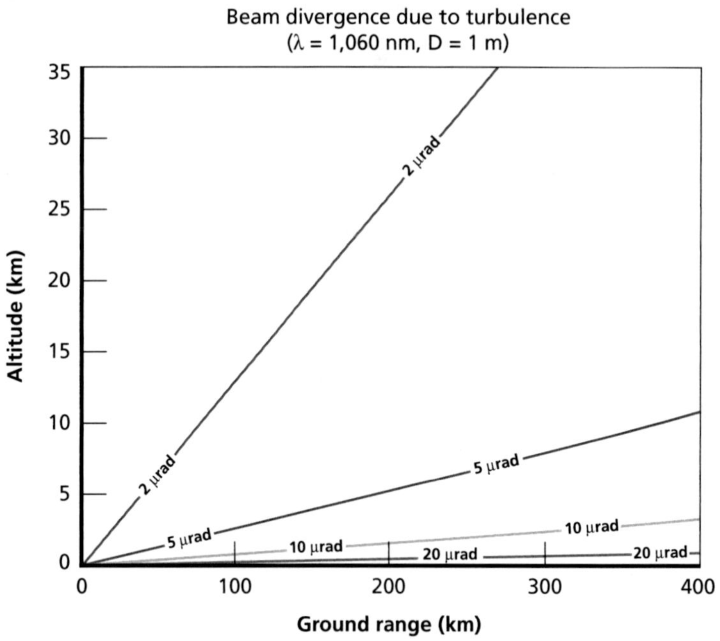

Beam divergence due to turbulence
(λ = 1,060 nm, D = 1 m)

RAND *TR898-C.2*

and near-horizontal approximations, where $\beta = 35°$, but the two approximations are of roughly the same magnitude where they meet. If the laser is situated at ground level, $\sigma_{turbulence}$ ranges from about 2 µrad on near-zenith transmission paths up to 10 µrad or higher if it is transmitting on a near-horizontal path to a low-altitude UAV.

If, instead, the laser is mounted on an airborne platform at an altitude of 40,000 ft (12.2 km), $\sigma_{turbulence}$ is reduced by about an order of magnitude, as shown in Figure C.3.

Jitter

Jitter is caused by any shaking, vibration, or other motion of the laser and optics. Cook (2004) suggested an allowance of σ_{jitter} = 4 µrad for total system jitter in a high-energy laser, while Kare (2004) budgeted for σ_{jitter} = 6 µrad.

If the laser is mounted on a ship or an aircraft, the motion of the vehicle could contribute to jitter. However, isolating the laser on an inertially stabilized platform can reduce jitter to less than 10 µrad, even for vehicles with highly dynamic motion, such as a maneuvering aircraft or a ship in a high sea state (Masten, 2008).

In the following calculations, I assume that σ_{jitter} = 6 µrad for ground-based laser transmitters, but σ_{jitter} = 10 µrad for airborne lasers.

Thermal Blooming

Thermal blooming occurs because the laser beam heats the column of air in the beam path, creating a region of lower air density in the center of the beam, which then acts as a negative

Figure C.3
Beam Divergence Caused by Turbulence for an Airborne Laser at an Altitude of 12.2 km (40,000 ft)

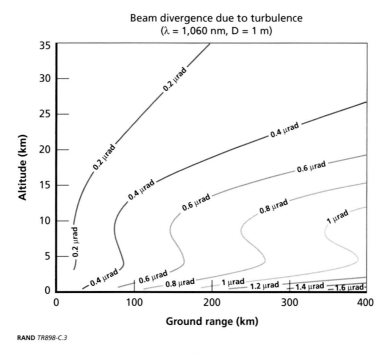

RAND *TR898-C.3*

lens, defocusing the beam. If $\sigma_L^2 = \sigma_{\text{diffraction}}^2 + \sigma_{\text{turbulence}}^2 + \sigma_{\text{jitter}}^2$, an estimate of the variance due to thermal blooming is (Tyson and Ulrich, 1993)

$$\sigma_{bloom}^2 = 0.0143\sigma_L^2\Psi^{1.178} \qquad (C.9)$$

where

$$\Psi = \frac{N_B N_F}{N_Q N_A}\exp\left(\frac{-\varepsilon L}{2}\right)\ln\left[\frac{(N_A+1)}{(N_A N_C - N_C^2)}\right]$$

$$N_F = \frac{\pi w_0^2}{2\lambda L}$$

$$N_Q = B\frac{\sigma_L}{\sigma_{\text{diffraction}}}$$

$$N_C = 2.5\sigma_L\frac{L}{w_0}$$

$$N_A = \left(1+6.24\frac{\sigma_L^2 L^2}{w_0^2}\right)^{1/2}$$

$$N_B = 1.84\times10^{-8}\frac{P\alpha_{\text{absorption}}L^2}{w_0^3 v}$$

and v is the wind velocity, including relative wind velocity due to motion of the laser beam. Solar-powered UAVs such as Helios and Zephyr have a minimum airspeed at altitude of 20 m/s or more, so in the calculations of Chapter Three, I assumed a conservative value of $v = 10$ m/s for relative wind due to motion of the laser beam. For the laser powers considered here (up to 25 kW), thermal blooming is not a significant issue; it is generally less than 1 μrad.

References

Airborne Laser System Program Office, "Airborne Laser (YAL-1A) Fact Sheet," Kirtland Air Force Base, N.M., 2003.

Aldair, P. L., Zheng Chen, and M. F. Rose, "Photoelectric Conversion Efficiencies for InGaAs Photovoltaic Cells Illuminated by Composite Selective Emitters," in *Proceedings of the 31st Intersociety Energy Conversion Engineering Conference*, Vol. 2, 1996, pp. 1018–1022.

Blackwell, Tim, "Recent Demonstrations of Laser Power Beaming at DFRC and MSFC," in *Third International Symposium on Beamed Energy Propulsion*, ISSN 0094-243X, No. 766, 2004, pp. 73–85.

Boehm, Albert R., and James H. Willand, "Probability of Cloud Statistics (PCloudS)" (computer program), Version 1.0, Albuquerque, N.M.: U.S. Air Force Phillips Laboratories, 1995.

Brown, Gerald V., Albert F. Kascak, Ben Ebihara, Dexter Johnson, Benjamin Choi, Mark Siebert, and Carl Buccieri, *NASA Glenn Research Center Program in High Power Density Motors for Aeropropulsion*, Cleveland, Oh.: NASA, John H. Glenn Research Center at Lewis Field, NASA/TM-2005-213800, December 2005.

Burle Industries, *Electro Optics Handbook*, Lancaster, Pa., 1974.

Cathcart, Michael, *MILPAS High Energy Lasers*, Atlanta, Ga.: Georgia Institute of Technology, 2007.

Cocconi, Alan, *AC Propulsion's Solar Electric Powered SoLong UAV*, San Dimas, Calif.: AC Propulsion, 2005.

Cook, Joung R., "Atmospheric Propagation of High Energy Lasers and Applications," in *Third International Symposium on Beamed Energy Propulsion*, ISSN 0094-243X, No. 766, 2004, pp. 58–72.

Defense Advanced Research Projects Agency (DARPA), "Vulture II Broad Agency Announcement (BAA)," Arlington, Va.: Defense Advanced Research Projects Agency, DARPA-BAA-10-04, October 22, 2009.

Friedman, Ed, and John Lester Miller, *Photonics Rules of Thumb: Optics, Electro-Optics, Fiber Optics, and Lasers*, McGraw Hill Professional, 2003.

Giesen, Adolf, and Jochen Speiser, "Fifteen Years of Work on Thin-Disk Lasers: Results and Scaling Laws," *IEEE Journal of Selected Topics in Quantum Mechanics*, Vol. 13, No. 3, May/June 2007, pp. 598–607.

Greer, Donald, Phil Hamory, Keith Kraker, and Mark Drela, *Design and Predictions for a High-Altitude (Low-Reynolds-Number) Aerodynamic Flight Experiment*, Edwards Air Force Base, Calif.: NASA Dryden Flight Research Center, NASA/TM-1999-206579, July 1999.

Gregerson, Chris, John Bangert, and Fred Pappalardi, "Celestial Augmentation of Inertial Navigation Systems: A Robust Navigation Alternative," U.S. Naval Observatory/Space and Naval Warfare Systems Command, USNO/SPAWAR white paper, n.d.

Hecht, Jeff, "Photonic Power Delivery: Photonic Power Conversion Delivers Power Via Laser Beams," *Laser Focus World*, January 2006, pp. 113–117.

IPG, "High Power Fiber Lasers for Industrial Applications," 2008.

Jane's Unmanned Aerial Vehicles and Targets, "AeroVironment Centurion/Helios," November 10, 2006.

———, "QinetiQ Zephyr," January 30, 2007.

JDSU, "Photonic Power Module," n.d. As of November 21, 2010:
http://www.jdsu.com/products/photovoltaics/products/a-z-product-list/photonic-power-module.html

Kare, Jordin T., "Modular Laser Launch Architecture: Analysis and Beam Module Design," Seattle, Wash.: Kare Technical Consulting, Final Report, USRA Subcontract Agreement No. 07605-003-015, April 30, 2004, revised May 18, 2004.

_____, "Modular Laser Options for HX Laser Launch," in *Beamed Energy Propulsion: Third International Symposium on Beamed Energy Propulsion*, American Institute of Physics (AIP) Conference Proceedings, Vol. 766, April 27, 2005, pp. 128–139.

Kobayashi, Ken, *Development of a Balloon-Borne Hard X-Ray Spectrometer for Solar Flare Observation*, master's thesis, University of Tokyo, Department of Astronomy, 2000.

Krupke, W. F., R. J. Beach, V. K. Kanz, S. A. Payne, and J. T. Early, "New Class of CW High-Power Diode-Pumped Alkali Lasers (DPALs)," in *High-Power Laser Ablation 2004*, Livermore, Calif.: Lawrence Livermore National Laboratory, UCRL-PROC-203398, 2004.

Krupke, W. F., R. J. Beach, S. A. Payne, V. K. Kanz, and J. T. Early, "DPAL: A New Class of Lasers for CW Power Beaming at Ideal Photovoltaic Cell Wavelengths," in *2nd International Symposium on Beamed Energy Propulsion*, Melville, N.Y., UCRL-CONF-155610, 2003.

Landis, Geoffrey, "Photovoltaic Receivers for Laser Beamed Power in Space," *Journal of Propulsion and Power*, Vol. 9, No. 1, 1993, pp. 105–112.

Masten, Michael K., "Inertially Stabilized Platforms for Optical Imaging Systems," *IEEE Control Systems Magazine*, Vol. 28, No. 1, 2008, pp. 47–64.

Millard, Douglas, "QinetiQ's Zephyr UAV Flies for Three and a Half Days to Set Unofficial World Record for Longest Duration Unmanned Flight," August 24, 2008. As of November 26, 2010: http://qinetiq.com/home/newsroom/news_releases_homepage/2008/3rd_quarter/qinetiq_s_zephyr_uav.html

Mohring, H. D., D. Stellbogen, R. Schäffler, S. Oelting, R. Gegenwart, P. Konttinen, T. Carlsson, M. Cendagorta, and W. Herrmann, "Outdoor Performance of Polycrystalline Thin Film PV Modules in Different European Climates," in *Proceedings of the 19th European Photovoltaic Solar Energy Conference*, Paris, France, 2004, pp. 2098–2101.

Nagashima, Tomonori, Koji Hokoi, Kenichi Okumura, and Masafumi Yamaguchi, "A Germanium Back Contact Type Cell for Thermophotovoltaics," in *Conference Record of the Thirty-First IEEE Photovoltaic Specialists Conference*, Orlando, Fla., 2005, pp. 659–662.

National Aeronautics and Space Administration (NASA), website, n.d. As of November 26, 2010: http://www.nasa.gov

Noll, Thomas E., John M. Brown, Marla E. Perez-Davis, Stephen D. Ishmael, Geary C. Tiffany, and Matthew Gaier, *Investigation of the Helios Prototype Aircraft Mishap: Vol. I, Mishap Report*, Hampton, Va.: NASA Langley Research Center, January 2004.

Noth, André, *Design of Solar Powered Airplanes for Continuous Flight*, doctoral dissertation, ETH Zurich, Switzerland, 2008a.

———, *History of Solar Flight*, ETH Zurich, Switzerland: Autonomous Systems Lab, Swiss Federal Institute of Technology Zürich, 2008b.

Noth, André, Roland Siegwart, and W. Engel, *Design of Solar Powered Airplanes for Continuous Flight*, ETH Zurich, Switzerland: Autonomous Systems Lab, Swiss Federal Institute of Technology Zürich, December 2006.

Olsen, Larry C., Wenhua Lei, F. William Addis, William N. Shafarman, Miguel A. Contreras, and Kannan Ramanathan, "High Efficiency CIGS and CIS Cells with CVD ZnO Buffer Layers," in *26th IEEE Photovoltaic Specialists Conference*, Anaheim, Calif., 1997, pp. 363–366.

Page, Lewis, "UK's Zephyr Robo Sun-Plane in Record-Buster 2-Week Flight," *The Register*, July 23, 2010. As of November 26, 2010: http://www.theregister.co.uk/2010/07/23/zephyr_down_successful/

Page, R. H., C. D. Boley, A. M. Rubenchik, and R. J. Beach, "Diode-Pumped Alkali Vapor Lasers—A New Pathway to High Beam Quality at High Average Power," in *Solid State and Diode Laser Technology Review*, UCRL-PROC-212117, 2005.

Posthuma, Niels E., Johan van der Heide, Giovanni Flamand, and Jozef Poortmans, "Emitter Formation and Contact Realization by Diffusion for Germanium Photovoltaic Devices," *IEEE Transactions on Electron Devices*, Vol. 54, No. 5, May 2007, pp. 1210–1215.

Powalla, M., B. Dimmler, and K. H. Groß, "CIS Thin-Film Solar Modules—An Example of Remarkable Progress in PV," in *Proceedings of the 20th European Photovoltaic Solar Energy Conference*, Barcelona, Spain, June 6–10, 2005, pp. 1689–1694.

Selinger, Marc, "Boeing Fires New Thin-Disk Laser, Achieving Solid-State Laser Milestone," June 3, 2008. As of November 26, 2010:
http://www.boeing.com/news/releases/2008/q2/080603a_nr.html

Spectrolab. "CDO-100 Concentrator Photovoltaic Cell Datasheet," updated April 24, 2008. As of November 26, 2010:
http://www.spectrolab.com/DataSheets/TerCel/C1MJ_CDO-100.pdf

Tesla Motors, 2008. As of November 26, 2010:
http://www.teslamotors.com

Thornton, Earl Arthur, *Thermal Structures for Aerospace Applications*, Reston, Va.: American Institute of Aeronautics and Astronautics, AIAA Education Series, 1996.

TRUMPF Group, "Lasers: Solve Every Task Perfectly," 2008.

Tyson, Robert K., and Peter B. Ulrich, "Adaptive Optics," in Stanley R. Robinson (ed.), *The Infrared & Electro-Optical Systems Handbook: Vol. 8, Emerging Systems and Technologies*, Bellingham, Wa.: SPIE, 1993, pp. 165–237.

Walker, Jan, "DARPA Chooses Contractors for Vulture Program," April 21, 2008. As of November 26, 2010:
http://www.darpa.mil/news/2008/vulture.pdf

Wojtczuk, Steven J., "Long-Wavelength Laser Power Converters for Optical Fibers," in *26th IEEE Photovoltaic Specialists Conference*, Anaheim, Calif.: 1997, pp. 971–974.

Woods, Lawrence M., Rosine Ribelin, and Joseph H. Armstrong, "Next-Generation Thin-Film Photovoltaics," *IEEE Aerospace and Electronics Systems Magazine*, Vol. 22, No. 10, October 2007, pp. 20–24.

Zhdanov, B. V., J. Sell, and R. J. Knize, "Multiple Laser Diode Array Pumped Cs Laser with 48W Output Power," *Electronics Letters*, Vol. 44, No. 9, April 24, 2008, pp. 582–583.